Indigenous Peoples' Governance of Land and Protected Territories in the Arctic

Thora Martina Herrmann • Thibault Martin
Editors

Indigenous Peoples' Governance of Land and Protected Territories in the Arctic

Springer

Editors
Thora Martina Herrmann
Département de géographie
Université de Montréal
Montréal, QC, Canada

Thibault Martin
Université du Québec en Outaouais
Gatineau, QC, Canada

ISBN 978-3-319-79723-6 ISBN 978-3-319-25035-9 (eBook)
DOI 10.1007/978-3-319-25035-9

Springer Cham Heidelberg New York Dordrecht London

Softcover reprint of the hardcover 1st edition 2016

Printed on acid-free paper

Springer International Publishing AG Switzerland is part of Springer Science+Business Media (www.springer.com)

Preface

The Arctic encompasses nearly 30 million km^2 of marine and terrestrial ecosystems and is home to many indigenous societies, each having its own culture, traditions and way of life, who have developed long and enduring relationships with the natural environment through knowledge systems and practices. Arctic territories are since two decades the object of major initiatives aiming to protect the environment. As a result, it is in the Arctic that we find the greatest concentration of large National Parks.

Yet, the accelerating and interrelating forces of climate and socio-environmental changes are altering fragile Arctic ecosystems more rapidly than any other area of planet Earth; they are also leading to profound socioeconomic and societal-cultural transformations in the Arctic. To maximise the benefits of protected areas to people, it is important to have governance systems in place that engage local communities to give them a voice in the protection of values important to them. It is now a priority to explore the adequacy of current governance arrangements of protected areas in the Arctic, to innovate these systems or to create new ones to address the challenges and the opportunities arising from the important and rapid changes in a way that maintains sustainable use of Arctic natural resources and enhances the well-being of the Arctic communities living in and around protected areas. Arctic indigenous, aboriginal peoples, with their knowledge and approaches, are playing increasingly important roles in responding to needs for Arctic governance in an era of transformative change.

In several countries like Canada, Alaska, Norway and Sweden, aboriginal populations are now contributing to the governance of these protected areas. For example, in Canada, an Aboriginal Affairs Secretariat was created in 1999 within Parks Canada (the national institution responsible for the creation and administration of protected areas) to facilitate the participation of aboriginal people in the governance of Canada's natural and cultural heritage places. Since they are increasingly involved in the planning and governance of protected areas, aboriginal communities take advantage of this situation to bring forwards new initiatives that aim at promoting their cultural heritage and that demonstrate the relationship that exists between their territory and their way of life.

In this collective multidisciplinary book, we will frame critical questions, analyse key issues and draw the picture of the different approaches adopted by the Arctic States to include (or not) aboriginal people in governance of the protected areas.

Thus, *Indigenous Peoples' Governance of Land and Protected Territories in the Arctic* brings together 34 authors from various disciplines (e.g. ethnology, law, geography, history, archaeology and arts) and backgrounds (indigenous peoples, scientists and researchers and members of the policy community) to (i) document indigenous approaches to governance of land and protected areas in the Arctic at the local, regional and international level; (ii) explore new territorial governance models that are emerging as part of the indigenous governance within Arctic States, provinces, territories and regions; (iii) analyse and document the recognition or lack of indigenous rights regarding self-determination and local control relevant to the Arctic; and (iv) examine how traditional decision-making arrangements and practices can be brought together with governments in the process of good governance of land and protected territories.

Indigenous Peoples' Governance of Land and Protected Territories in the Arctic adopts a multidisciplinary approach with contributions from different perspectives and world views, from various geographical scales and governance levels. Contributions are cross-regional and based on case studies in 5 Arctic countries (Finland, Norway, Canada, USA (Alaska), Denmark (Greenland)). Our book contains a total of 12 chapters.

Modern-day land claim agreements and protected area agreements can create an opportunity to help reconcile the interests and world views of aboriginal and non-aboriginal societies. Authors in the first three chapters are analysing this issue. Thus, Jacobson, Manseau, Mouland et al., in the first chapter of our volume, offer an interesting analysis of the Auyuittuq National Park established in 2001 under the Nunavut Land Claims Agreement (NLCA). Park governance is a cooperative management in accordance with the NLCA, the Inuit Impact and Benefits Agreement and the National Parks Act. In practice, this has been a transformation in organisation from informing/liaising to active participation/engagement of adjacent communities. In fall 2012, the authors met with elders, park staff and Field Unit staff to explore Inuit aspirations for the park and to examine the ways in which the objective of knowledge gathering and sharing was being applied as part of management operations. Their chapter explores Pangnirtung Inuit perspectives on the park and identifies issues in the application of indigenous knowledge as part of its ongoing management. The authors contrast this against shifts in processes for implementing management, through analysis of policy and planning negotiations. They explore the implications of a western management system for Inuit directly involved in day-to-day operations of park management.

Taking the case of the Finnmark Estate, Josefsen, Søreng and Selle, in chapter "Regional governance and indigenous rights in Norway: The Finnmark Estate case", investigate how a regional political and administrative co-management reform in Norway, established to arrange for dialogue and cooperation between the indigenous Sámi and the state, faced resistance from the local population. The Finnmark Estate established in 2006 for implementation of indigenous rights in the management of land and natural resources operates in areas where the population is ethnically mixed and the indigenous people are a minority. The authors explore the public's attitude towards this new regional governance body that shall secure indigenous rights along with other obligations and ask compelling questions: Has the opinion changed its positions since 2006 so that we now find less deep conflict? Under what conditions could a co-management structure, which includes regional government and indigenous representatives, gain public legitimacy? The authors highlight the complex interplay between legally adopted indigenous rights which are formalised in the regional governing system and the informal norms and barriers for change institutionalised in the surrounding society.

In three protected area and cooperative management agreements – the Saoyú-ʔehdacho Agreement (2008), the Tuktut Nogait Agreement (1998) and the Gwaii Haanas Agreement (1993) – aboriginal authorities and the government of Canada have agreed to use the aboriginal concept and practice of consensus decision-making in their cooperative management of the vast protected land pursuant to these agreements. Tom Nesbitt, in chapter "Increasing cooperation and advancing reconciliation in the cooperative management of protected areas in Canada's north", discusses these agreements through the lenses of reconciliation, consensus decision-making and cooperative management of protected areas.

Numerous issues facing individual protected area agencies reach beyond park and national boundaries and also affect neighbouring protected areas or countries. Issues of common concern provide opportunities to work collaboratively on a continental scale, such as the transboundary protected area network in the Barents Euro-Arctic Region, or on a regional scale to improve protected area management. Lemelin, Johnston, Lough et al. in chapter "Two parks, one vision – collaborative management approaches to transboundary protected areas in northern Canada: Tongait KakKasuangita SilakKijapvinga/Torngat Mountains National Park, Nunatsiavut and le Parc national Kuururjuaq Nunavik", address this question and study how collaborative management strategies have been implemented at the regional level by applying the Indigenous Stewardship Model to the Tongait KakKasuangita SilakKijapvinga/Torngat Mountains National Park (TMNP) and Kuururjuaq Parc National (KPN), Canada's newest polar transboundary protected area along the Labrador Peninsula of Northern Quebec and Newfoundland and Labrador. The adaptive management approach used in this transboundary protected area nurtures regionally based approaches to protected area management and promotes regional collaborative developments. These regional initiatives are facilitated through an all Inuit Co-operative Management Board for the TMNP and a harmonisation committee overseeing the management of the KPN. Their analysis shows that although the mandate of each park committee is to provide advice and guidance for the management of their respective parks, each has also become an important forum for facilitating more regionally based governance and management approaches through protected areas.

Developmental pressures are increasingly exerted on the Arctic and challenging the way of life of its many indigenous peoples. Against this background, there seems to be a growing need for measures to protect special areas in the North. This is the focus of chapter "Conceptual and institutional frameworks for protected areas, and the status of indigenous involvement: considerations for the Bering Strait Region of Alaska", by Raymond-Yakoubian. Looking at the Bering Strait region of Alaska, she argues that, from the perspective of indigenous people, additional protected areas – particularly in relation to the marine ecosystems – are needed in this region in light of the rapid and dramatic changes, both climate- and development-related facing the area, such as commercial fishing, increasing marine traffic, climate change and resource development. The author reviews some of the existing protections that are in place and the status of indigenous involvement in them and suggests paths for extending beyond typical western understandings of the nature, to include indigenous residents of the region in the development, creation and maintenance of protected areas. If agencies and governments take the time to develop relationships with tribes and tribal members in the Bering Strait region, it will not only enhance support for protections, but collaboration with tribes and their traditional knowledge base will lead to better decision-making regarding the need for protected areas and determining what form protections could most effectively take.

Ecological, cultural, spiritual, aesthetic, recreational and economic values of protected areas across the Arctic are undoubtedly immense, and the following three Chapters are exploring these.

The interaction of spiritual elements, indigenous peoples and protected areas in the Arctic is considered in chapter "Protecting the 'Caribou Heaven': a sacred site of the Naskapi and protected area establishment in Nunavik, Canada", which puts the emphasis on Sacred Natural Sites – the world's oldest protected places in the Arctic. Mameamskum, Herrmann and Füleki look at the recognition and conservation of sacred natural sites located within the boundaries of legally established protected areas, in Nunavik (Canada). Taking the case of the Caribou Heaven, a spiritually and culturally important site for the Naskapi First Nation situated within the limits of this Kuururjuaq National Park, the authors describe how Naskapi ecological knowledge was used to designate this sacred site as an area of extreme protection within the Park, to ensure its preservation and integrity. They describe how cultural and spiritual values have formed the basis of indigenous management models of nature conservation in this park.

Tommasini in the following chapter "The governance of protected areas in Greenland: the Resource National Park among conservation and exploitation" presents an example of protected area costs and benefits to indigenous well-being, specifically within the sectors of tourism. She focuses on the world's largest and most northerly protected area – the Northeast Greenland National Park. This park is strictly regulated for its access and allowed activities, e.g. recreational and outdoor activities are not authorised, and permission is needed, except for the population living adjacent to the Park, to be in the region, but other activities, for instance, mineral pits, are allowed. She explores the role of the local population in the governance of this national park seen from the local point of view as a resource for the socioeconomic revitalisation of the adjacent community of Ittoqqortoormiit.

Ecological concerns for wildlife species lie often at the heart of the protective impulse that drives the establishment of protected areas in the Arctic. Policies relating to wildlife co-management can create opportunities but are also often accompanied by conflicts between indigenous and nonindigenous attitudes towards resource use. In Canada protected areas cover over 250,000 ha within the polar bear range. The co-management of polar bears between scientists and the Inuit in Nunavut has been fraught with tension. Vaudry, in chapter "Conflicting understandings in polar bear co-management in the Inuit Nunangat: enacting Inuit knowledge and identity", explores the Inuit's perspective by highlighting where the bear fits within Inuit cosmology and how it influences their relationships with the animal, with respect to hunting. Since 2005, environmentalists have tried to ban polar bear hunting on an international scale and to get the animal put on the list of species threatened with extinction. This has had a major impact on already fragile northern economies, as it discourages sport hunting, which many Inuit count on for needed income. The author urges that the debate surrounding the regulation of polar bear hunting in the Canadian Arctic cannot be settled without the Inuit point of view being considered or without Inuit being part of the decision-making process.

Incentive towards protected area reconciliation and rights-based reform in protected area design and governance comes also from greater attention to international legal mechanism. Of particular relevance are the International Labour Organization (ILO) Convention No. 169 concerning indigenous and tribal peoples in independent countries and the United Nations Declaration on the Rights of Indigenous Peoples (DRIPS). The following three chapters discuss aspects of indigenous rights and legal instruments regarding self-determination and local control relevant to the governance of lands and protected areas in the Arctic. Thus, Martin, in chapter "Beyond the protection of the land, national parks in the Canadian Arctic: a way to actualized and institutionalized aboriginal cultures in the global", offers a comprehensive approach to aboriginal governance of protected areas by examining national parks located in the Canadian Arctic, mainly in Inuit territory. Whereas Parks Canada long excluded aboriginals from the governance processes of protected areas, today indigenous peoples, as signatories to political agreements, have legal tools to access equality on park management councils. Yet, notwithstanding these legal progresses, not all co-management models give the same space to indigenous communities, and, as this chapter reveals, the management councils express more or less satisfaction with the management model in which they are involved. Martin underlines co-jurisdiction as the form of co-management favoured by aboriginal people as it creates legal conditions for an egalitarian partnership based on recognition of their land rights and knowledge.

In chapter "Recognition of indigenous lands through the Norwegian 2005 Finnmark Act: an important example for other countries with indigenous people?", Oyvind examines the commitment to identify and recognise indigenous people's lands and natural resources in relation to the indigenous Sámi in the Nordic countries. This commitment applies in particular to Norway, which is the only country with a Sámi population who has ratified the ILO Convention. The commitments imposed to Norway thus raise several key issues regarding identification of indigenous people's lands, including to what extent the Sámi laws and customs have significance as legal sources in such processes and how the state must involve the indigenous party in the process.

In international law, a fundamental shift is currently occurring in state-indigenous relations, which can be seen as culminating in the adoption of the UN Declaration on the Rights of Indigenous Peoples and at its endorsement of the rights of indigenous peoples to self-determination and a free, prior and informed consent (FPIC) in the decisions that concern them. Heinämäki, in chapter "Global context – Arctic importance: free, prior and informed consent, a new paradigm in international law related to indigenous peoples", analyses FPIC in the light of developments that have prepared and pushed states to slowly accept that indigenous peoples must be recognised as serious actors and as "partners" with and within the nation states. When implemented, the right to FPIC can have positive effects on the important issues such as indigenous peoples' land use and governance. This is of a particular importance in the Arctic that is the homeland for a great number of indigenous peoples.

Positioning Arctic Canada as central to its political platform, the current federal government has set itself up to be one of Canada's most northern focused federal regimes in decades. Sinclair in the last chapter "Untouched and uninhabited: conflicting Canadian rhetoric on the protection of the environment and advancing northern economies" offers an interesting examination of the ways that the Prime Minister's policy speeches portray and frame the Canadian Arctic's environment and land use and the extent to which these statements incorporate broader ideas and premises about Arctic Canada. In particular, this involves an examination of the effects of policy portrayals in speeches of Arctic Canada as an untouched and uninhabited wilderness. This feeds into a tension between resource extraction in "isolated" areas as justifiable and the impulse to protect pristine places and the interests of northern residents, leading to a possible exclusion of northern, and in particular Inuit, priorities.

The 12 chapters taken as a whole provide strong and compelling evidence for the recognition of the rights of indigenous peoples, equitable cost and benefit-sharing and new forms of governance to protected area management in the Arctic. Together, they:

- Provide an interdisciplinary overview of key issues regarding indigenous peoples and governance of land and protected areas in Arctic regions
- Explore new territorial governance models that are emerging as part of the indigenous governance within Arctic states, provinces, territories and regions
- Discuss aspects of indigenous rights regarding self-determination and local control relevant to the Arctic
- Forge a new understanding of how traditional decision-making arrangements and practices can be brought together with governments in the process of good governance of land and protected territories in the Arctic at the local, regional and international level
- Identify key principles, lessons learnt, that are useful to address issues of Arctic governance of land and protected territories today and that could be relevant for future governance arrangements

Through the diverse contributions, our book *Indigenous Peoples' Governance of Land and Protected Territories in the Arctic* aims to offer the right balance between locally, regionally and internationally focused and thematic chapters. We adopted a multidisciplinary and multi-scale approach and hope this approach makes the book enjoyable to read. Producing this book has been an interesting and valuable experience for us. We have learnt from each other's practices and methods and have broadened our perspectives.

Indigenous Peoples' Governance of Land and Protected Territories in the Arctic is aimed at environmental and social scientists, local communities, policymakers and planners. We hope that scientists in environmental conservation, cultural geography, sociology, political sciences and law will find this book insightful. We particularly hope that practitioners working in the area of protected area planning and management will find our book useful. Conservation professionals might also benefit much from the comprehensive case studies and the experience they contain on how to establish co-management arrangements of specific areas in a way that is adequate to the well-being of Arctic communities. It is our wish that our book makes a valuable contribution to the emerging literature on aboriginal governance, especially in that it offers a regional perspective and direct experience and case studies from Arctic aboriginal communities.

We are now in the Second International Decade of the World's Indigenous People (2005–2015) with the theme "Partnership for action and dignity". As such, we like our book to be a concrete contribution towards a more equitable and effective governance and management of protected areas in the Arctic. These areas are an important tool for the conservation of biocultural diversity in the North, a cornerstone of sustainable development strategies in Arctic regions and a means for preserving the rich Arctic heritage of the land and local cultures to future generations.

Département de géographie
Université de Montréal
Montréal, QC, Canada
Thora Martina Herrmann

Université du Québec en Outaouais
Gatineau, QC, Canada
Thibault Martin

Contents

Part I
Modern-Day Land Claim Agreements and Governance Agreements in the Arctic – An Opportunity to Reconcile the Interests and Worldviews of Aboriginal and Non-Aboriginal Societies?

Co-operative Management of Auyuittuq National Park: Moving Towards Greater Emphasis and Recognition of Indigenous Aspirations for the Management of Their Lands

Chris Jacobson, Micheline Manseau, Gary Mouland, Amy Brown, Andrew Nakashuk, Billy Etooangat, Matthew Nakashuk, Delia Siivola, Leesee-Mary Kaki, Jaypetee Kapik, Manasa Evic, Abraham Kennianak, and Davidee Koonelieusee

Abstract Auyuittuq National Park (Nunavut, Canada) was first established in 1976 as a national park reserve under the National Parks Act of Canada. It was subsequently established as a national park in 2001 pursuant to the Nunavut Land Claims Agreement (NLCA). Park governance is currently a co-operative management framework in accordance with the NLCA, the Inuit Impact and Benefits Agreement and the National Parks Act. In practice, this has been a transformation in organisation

C. Jacobson (✉)
Sustainability Research Centre, University of the Sunshine Coast, Sippy Downs, Australia
e-mail: cjacobso@usc.edu.au

M. Manseau
Natural Resources Institute, University of Manitoba,
70 Dysart Road, Winnipeg, MB R3T 2N2, Canada

Office of the Chief Ecosystem Scientist, Parks Canada,
30 Victoria Street, Gatineau, QC J8X 0B3, Canada
e-mail: micheline.manseau@pc.gc.ca

G. Mouland
Office of the Chief Ecosystem Scientist, Parks Canada,
30 Victoria Street, Gatineau, QC J8X 0B3, Canada

A. Brown
Natural Resources Institute, University of Manitoba,
70 Dysart Road, Winnipeg, MB R3T 2N2, Canada

A. Nakashuk • L.-M. Kaki • J. Kapik • M. Evic • A. Kennianak • D. Koonelieusee
Care of Auyuittuq National Park, Parks Canada Agency,
P.O. Box, C.P. 353, Pangnirtung, NT X0A 0R0, Canada

B. Etooangat • M. Nakashuk • D. Siivola
Auyuittuq National Park, Parks Canada Agency,
P.O. Box, C.P. 353, Pangnirtung, NT X0A 0R0, Canada

T.M. Herrmann, T. Martin (eds.), *Indigenous Peoples' Governance of Land and Protected Territories in the Arctic*, DOI 10.1007/978-3-319-25035-9_1

from informing and liaising to active participation and engagement of adjacent communities. Although a management plan was recently developed by the Joint Park Management Committee, examination of Inuit aspirations for the Park has been limited. These aspirations form a basis for interpreting the effectiveness of the existing governance arrangements. Further, as an informal arrangement, Inuit Knowledge Working Groups have recently been established to also improve on engagement of the community; in sharing of knowledge and perspectives in support of park research and management activities. The success of this project also remains unexamined.

In the fall of 2012, we met with elders, park staff and Field Unit staff involved in management to explore Inuit aspirations for the park, and to examine the ways in which the objective of knowledge gathering and sharing was being applied as part of management operations, focussing on the community of Pangnirtung where the Park Office is located. A separate study focussed specifically on youth. This chapter explores Pangnirtumiut perspectives on the park, and identifies ongoing issues in the application of Indigenous knowledge as part of its ongoing management. We also contrast this against shifts in processes for implementing management, through analysis of policy and planning negotiations.

Inuit aspirations for the park include concerns for visitor safety, employment, protection of Aboriginal rights (including hunting), and an interest in sharing the history and use of the park area. Despite being called the Joint Park Management Committee, the committee plays more of an oversight role, ensuring key aspirations are appropriately addressed. Other initiatives, for example the Inuit Knowledge Working Groups as an opportunity for community engagement play an important role in sharing of knowledge more broadly, with a focus on the importance of being and remaining knowledgeable about the Land. We explore the implications of a western management system for Inuit directly involved in day to day operations of park management. This, and the perspective and opportunity for youth engagement capture the forward thinking aspirations of the community. In this sense, the park can be defined not as a physical boundary but a social one that captures in a snapshot a process of self-determined community change.

Keywords Protected areas management • Aboriginal knowledge • Co-operative management • Aboriginal land claim • Nunavut

1 Introduction

Co-operative management involves communities and governments formally sharing the management of the environment and the natural resources within it (McCay and Jentoft 1996). It has its roots in academic critique and practical solution-finding, regarding who ought to be involved in management, from a position of legitimising community rights and reconciling competing property claims (Plummer and

Fitzgibbon 2004; Carlsson and Berkes 2005). Co-operative management of natural resources is not new (Borrini-Feyerabend 1996); it is one of four recognised governance types for protected areas (Dudley 2008). It suggests an ideal for participatory management that enhances equity of indigenous groups, and helps to ensure appropriate distribution of benefits from conservation.

Globally, there has been an increase in co-operative management of protected areas and natural resources (especially wildlife), particularly as Indigenous peoples have settled grievances with colonial governments (e.g. Australian Institute of Aboriginal and Torres Strait Islander Services 2012). Although many authors have written about the benefits of co-operative management arrangements (e.g. Plummer and Fitzgibbon 2004) some challenges do arise from the different ideas and worldviews and ideas about how to conserve resources, appropriate representation and management processes, and in the application of Indigenous knowledge to management (Weitzner and Manseau 2001; Langdon et al. 2010). In this chapter, we explore the successes and challenges for co-operative management within the context of a modern land rights settlement.

Canada has over 25-years' experience with co-operative management of National Parks (Weitzner and Manseau 2001; Murray 2010), and the number of co-operatively managed protected areas is growing. In 2010, 68 % of lands were managed by Parks Canada under formal co-operative management agreements, reflecting a broader shift towards greater aboriginal engagement including changes to guiding principles and operating policies (Langdon et al. 2010). At time of publishing, 13 National Parks and National Park Reserves were co-operatively managed, and an additional three were in negotiation (Dave Murray, Parks Canada, April 2013, personal communication). Co-operative management arrangements in Canada's north have arisen during the settlement of land claims.

Auyuittuq National Park was declared a National Park Reserve in 1976. It covers 19,089 km^2 in central Baffin Island, lying largely within the Arctic Circle. The park incorporates land and marine areas, including the Penny Ice Cap (Parks Canada 2010) and abuts two communities, Qikiqtarjuaq in the north, and Pangnirtung in the south (Fig. 1), with populations of 520 and 1425 respectively (Statistics Canada 2012). The demographic of these communities is young (49 % and 56 % of the population [respectively] are under the age of 20), household sizes are larger than the Canadian average (3.1 and 3.7 respectively) with a predominance of Inuktitut spoken in the home (92 % and 88 % respectively, with neither English nor French spoken in 13 % and 26 % of cases) (Statistics Canada 2012). Like many Inuit communities, Pangnirtung existed first as a missionary outpost, with a slow shift between the 1930s and 1970s from semi-nomadic to more permanent settlement (Tester and Kulchyski 1994; Charles and Trott 2010). Since then, its population has grown rapidly. Pangnirtung is the location of the main Parks Canada Auyuittuq National Park Visitor Centre, although there is a smaller satellite office in Qikitarjuaq.

Co-operative management of Auyuittuq was negotiated as part of the Nunavut Land Claims Agreement (1993). National Parks are part of a system with the primary goal to protect and present natural areas of Canadian significance but they also provide economic and social benefits. In the case of Auyuittuq, those economic and

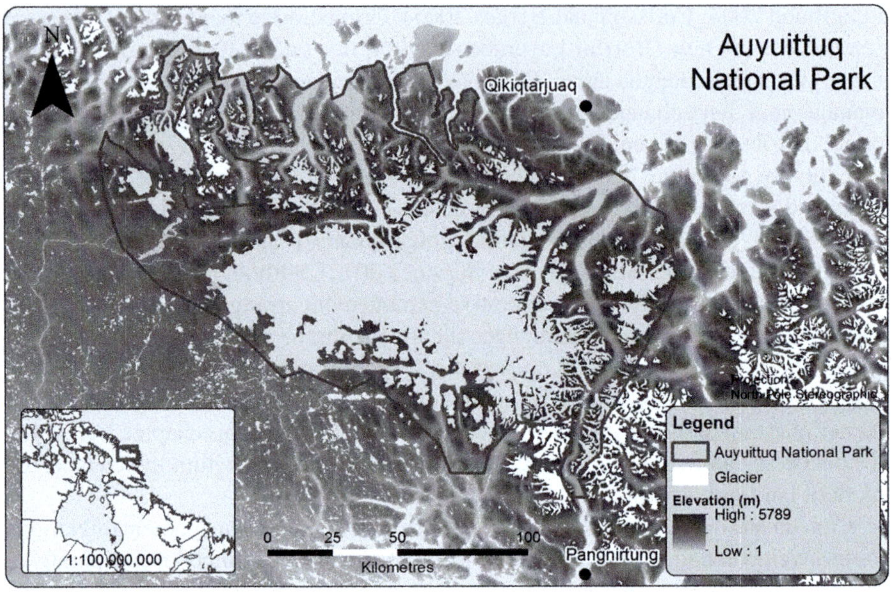

Fig. 1 Location of Auyuittuq National Park, Nunavut, Canada, and adjacent communities

social benefits, including a mechanism for Inuit to apply their knowledge and cultural understanding to park management, have been negotiated and specified in the Inuit Impact and Benefit Agreement for Auyuittuq, Quttinirpaaq and Sirmilik National Parks (IIBA 1999). Thus, a multi-faceted approach to management is necessary. The recentness of this co-operative management arrangement, and its embedding in land rights settlement, provides a contemporary example of co-operative management in the circumpolar north. It is important to note that Parks Canada uses the term co-operative management instead of co-management to describe the advisory nature of the management framework. Aboriginal groups work collaboratively with Parks Canada on the management of the parks, the co-operative management board advise the Minister on matters pertaining to the planning and operations of the park but the Minister retains the ability to make final decisions (Langdon et al. 2010).

2 Case Study Context

In 1993, the Nunavut Land Claims Agreement was signed, although Nunavut itself did not come into effect until 1999 (Mercer 2008). Central to the governance of the Nunavut Territory and to the ongoing governance of Auyuittuq are two key features: *Inuit Qaujimajatuqangit* (IQ), and co-operative management. Whilst the territory governs for all its residents, government acknowledges a set of principles based on IQ. Defined in many ways, IQ is "that which Inuit have long known" (Henderson

2009) and includes knowledge (in the Western sense), how people 'fit' with the environment, their ethical values and the way they behave towards it (Wenzel et al. 2008). IQ forms the underlying fabric of Nunavut government, including a set of principles for governance (e.g. *ajiiqatigiinniq* – decision making through consensus), an emphasis on Inuktitut (the native language), and a focus on activities that enable 'knowledge' to evolve (Pearce et al. 2011; Usher 2000; Mercer 2008). The Nunavut Land Claims Agreement provides for five Nunavut co-management boards (the Nunavut Wildlife Management Board, the Nunavut Impact Review Board, the Nunavut Planning Commission, the Surface Rights Tribunal and the Nunavut Water Board) that facilitate the incorporation of IQ into public resource management, through board membership appointments by Nunavut Tunngavik Incorporated, the territorial Inuit association (Legare 1996). The embedding of governance and co-operative management in IQ can be viewed as an attempt to support cultural pluralism and address colonial legacies through the empowerment of Inuit constituents in the Territory (Henderson 2009; Tester and Irniq 2008). The Nunavut Land Claims Agreement resulted in the Inuit Impact and Benefit Agreement for Auyuittuq, Quttinirpaaq and Sirmilik National Parks (IIBA 1999) that underpins the co-operative management of Auyuittuq, Sirmilik and Quttinirpaaq National Parks. Box 1 provides a summary of the objectives of the IIBA.

Box 1: Objectives of the IIBA and Principles and Objectives of the Park Planning Program

IIBA Objectives (Parks Canada and Qikiqtani Inuit Association 1999)

(a) to enhance cooperation between Inuit and the Government of Canada and to strengthen Inuit participation in the planning, management and operation of the Parks;
(b) to recognize that Inuit are an integral part of the ecosystems of the Parks;
(c) to honour the rights of Inuit in the Parks as set out in the Nunavut Land Claims Agreement and to promote greater awareness of these rights;
(d) to establish the Parks as part of a system of National Parks that are dedicated to the people of Canada for their benefit, education and enjoyment;
(e) to enhance the management of the Parks and the sustainable use of resources in the Parks by integrating Inuit knowledge, culture, and practices into the protection and conservation of the Parks and their resources, so as to leave the Parks unimpaired for future generations;
(f) to provide opportunities for Inuit in the adjacent communities to benefit from the establishment, planning, management and operation of the Parks; and
(g) to build a positive and effective relationship between the Parties to ensure that this IIBA is implemented with the spirit and intent with which it was negotiated.

Prior to the Nunavut Land Claim Agreement, Auyuittuq was a National Park Reserve with local advisory committees in both Pangnirtung and Qikiqtarjuaq, operating from 1982, with the purpose of advising the park manager on specific aspects of management (Parks Canada 1982). Inuit Impact and Benefit Agreement for Auyuittuq, Quttinirpaaq and Sirmilik National Parks (1999), an agreement between the Federal government and the Qikiqtani Inuit Association (one of three Regional Inuit Associations affiliated with Nunavut Tungavik Incorporated and representing Inuit from the Baffin region) provides guidance for the co-operative management of the park including establishment of two committees: the Joint Park Management Committee and beneath it, the Park Planning Team. The Inuit Knowledge Working Group exists as an informal advisory committee. Figure 2 provides a summary of the relationships.

Joint Park Management Committee (JPMC) membership consists of three appointments each from the Qikiqtani Inuit Association and the Federal Minister with responsibility for National Parks, although committee members represent the interests of all Canadians. The role of the JPMC (IIBA 1999, article 5) includes input into the management of outpost camps, carving stone removal, water licencing, protection of archaeological sites, planning, research, park promotion and information, park displays, exhibits and facilities, visitor access and use of the park,

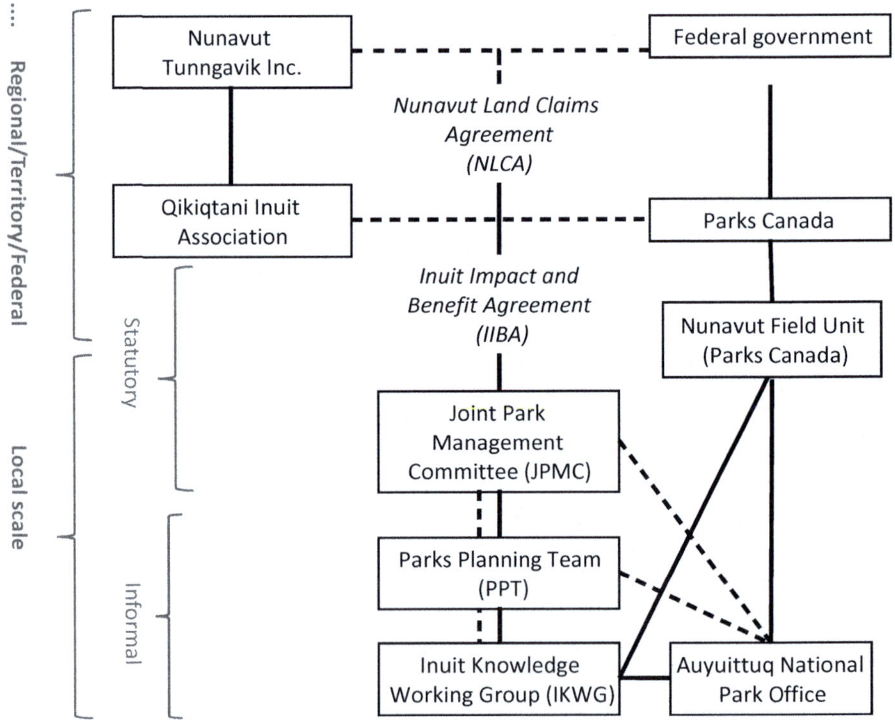

Fig. 2 Governance arrangements for Auyuittuq National Park of Canada

employment and training of Inuit employees, economic opportunities and changes to park boundaries and management planning (including management plan and business plan approval). The JPMC meets twice each year, with opportunity for teleconference contact as needed. The Nunavut Field Unit Superintendent (Head person for the National Parks in Nunavut) and the Qikiqtani Inuit Association have observer status to all meetings. Although the JPMC may provide advice to the Federal Minister for the Parks Canada Agency, the Minister may reject, accept or vary that advice at his or her discretion. The JPMC may also provide decisions to the Minister on specific matters outlined in the IIBA and the Minister may reject or accept those decisions as well. However, a rejection of a JPMC decision can only be made if the Minister determines that the decision is contrary to any one of several conditions specified in the IIBA. It is important to note that this has not happened as yet at Auyuittuq. The IIBA 1999 allows for annual JPMC funding, including funding for the JPMC secretariat. In addition to the Auyuittuq JPMC, a tripartite-JPMC (TRI-JPMC), where JPMC members from all three parks covered by the IIBA 1999 (Auyuittuq, Sirmilik and Quttinirpaaq) can meet has been supported by the Nunavut Field Unit.

Like the JPMC, the Parks Planning Team (PPT) consists of equal appointments from the Qikiqtani InuitAssociation and Parks Canada. The team is responsible for the development of a draft management plan for submission to the JPMC for review both prior to and after public consultations. Approval is then required by the Nunavut Wildlife Management Board, the Parks Superintendent and the relevant Federal Government Minister. The IIBA provides detailed information regarding the principles and objectives of the program, a number of which relate to Inuit Knowledge (see Box 2), the purpose of the park, and the process for plan development (including consultation), change, and review. The Auyuittuq National Park Management Plan was approved in 2010. The IIBA itself is subject to review every 7 years, as outlined in the Nunavut Land Claims Agreement (section 8.4.7). Thus, there is opportunity for negotiated adaptation and adjustment of procedures based on experiences of communities they serve.

Box 2: Principles and Objectives of the Park Planning Program (with Emphasis on Knowledge Underlined)

Principles of the Park Planning Program

(a) the Park is a part of ecosystems that extend beyond the boundaries of the Park;
(b) Inuit are an integral part of these ecosystems;
(c) the Management Plan will provide clear direction related to protecting, presenting and managing these ecosystems in the Park
(d) the Program will be carried out in a manner that is consistent with the National Parks Act, the NLCA, this Agreement, and Parks Canada policies;
(e) an effective Program requires the active cooperation and participation of both Inuit and the Government;

(continued)

Box 2: (continued)

(f) the Program will give equal consideration to scientific information and Inuit knowledge;
(g) park zoning and visitor use in the Management Plan will be consistent with maintenance of the ecological integrity of the Park and with Inuit access to and use of the lands and resources of the Park as provided for by the NLCA and this Agreement;
(h) ongoing basic and applied research is necessary in order to make responsible decisions for the planning, management and operation of the Park; and,
(i) Zones I and II, as defined and used in Article 8 of the NLCA, will constitute the predominant proportion of zoning in the Park.

Objectives of the Park Planning Program

(a) to protect the parts of the ecosystems in the Park in order to maintain the ecological integrity of the Park;
(b) to recognize the fact that Inuit are an integral part of the ecosystems of the Park;
(c) to respect and incorporate the knowledge, experience and practice of both Inuit men and women;
(d) to cooperate with Government agencies, departments, institutions of public government, non-government agencies and other levels of government to facilitate an ecosystem-based approach to the maintenance of the ecological integrity of the Park;
(e) to provide for opportunities for Inuit to participate in economic endeavours associated with the establishment, management and operation of the Park;
(f) to integrate the management and operation of the Park with activities and concerns of Inuit of the adjacent communities and region in a manner that assists the communities in the promotion and development of park-compatible regional tourism, including the pursuit of public or private partnerships;
(g) to provide visitors to the Park with appropriate opportunities to appreciate and understand the ecosystemic relationship between Inuit and the lands and resources found in the Park;
(h) to manage visitor access to and use of the lands and resources in the Park with a fundamental aspect being minimal interference with access to and use of the lands and resources in the Park by Inuit, as provided for in the NLCA and this Agreement;
(i) to respect and recognize the role of Inuit in the interpretation of place names and resources directly related to Inuit culture;
(j) to provide opportunities for public participation in the Program by means of public consultations at a local, regional or national level, as appropriate; and,
(k) any other objectives in the National Parks Act or Parks Canada policies.

In 2005, Parks Canada staff secured research funding for the Inuit Knowledge Working Group (IKWG) to work with the communities surrounding the national parks in Nunavut, with the primary goal of increasing "knowledge of the parks while allowing Parks Canada to gain a better appreciation and understanding of Inuit knowledge" (Inuit Knowledge Project – www.lecol-ck.ca/ik). This is part of Parks Canada's corporate culture change, where groups help to guide research projects and "create an environment of sharing and cooperation that has not only strengthened the level of understanding about the natural environment and cultural landscape but has also solidified long-term relationships" (Langdon et al. 2010: 230). Each group is composed of two Elders appointed by the Elders' Society, one member appointed by the Hunters and Trappers Organisation and one Youth member selected by other members (Inuit Knowledge Project – www.lecol-ck.ca/ik). The IWKG meet on a needs basis, but at least once a year and provide input into research priorities for the Park and advice on various research projects (e.g. climate change, sea ice safety travel, history) and operational programs including the Cultural Resource Value Statement, the Zoning Plan, the State of the Parks report, environmental monitoring and staff training. In this regard, the IKWG has a role in delivering the key principles and objectives of park management. According to the Management Plan, "the knowledge, experience and approaches developed in the project (IKWG) will be used by the staff of Nunavut to improve and strengthen Parks Canada's ecological integrity research and management programs" (Parks Canada 2010: 4).

Given the recentness of co-operative management of Auyuittuq National Park, we sought to understand the extent to which current governance arrangements allow for the attainment of economic, social and conservation aspirations Inuit have for the management of the Park (as emphasised in the IIBA 1999), and the extent to which arrangements meet the territorial emphasis on embedding park management in IQ. In fall 2012, we conducted a series of in depth interviews (13) and workshops (5) with staff, elders and other community members involved in the management of the park. These individuals have been involved in the Joint Park Management Committee (5), Park Planning Team (3), Inuit Knowledge Working Group (6) and as staff (7) at various stages since the National Park was established, and two were involved in advisory roles prior to the IIBA. Interviews and workshops discussed (i) their background and interest in the park, (ii) their perceptions of the roles and functioning of the governance groups they have been involved in, (iii) their experiences (positive and areas for improvement) of the groups, and (iv) the extent they feel Inuit Knowledge is being applied in park management, and if required, how this could be enhanced. Interviews and workshops were recorded and notes were taken before analysis using NVivo to identify key analytic themes. Additional background information was provided by parks staff at the Nunavut Field Unit in Iqaluit, and from documents on park planning and plan implementation.

3 Inuit Aspirations for Park Management and the Ability of Arrangements to Meet Them

All research participants were supportive of the park management and the structures and processes that enabled their involvement. The Joint Parks Management Committee, Parks Planning Team and the activities of the Inuit Knowledge Working Groups are significant moves forward in terms of Indigenous input into Parks Management (see also Langdon et al. 2010). Participants' key reasons for being drawn to participate in co-operative management included:

- knowledge – the use of *Inuit Qaujimajatuqangit* , learning from one another, and a chance to pass on knowledge;
- well-being – the safety of visitors in an extreme environment, the cleanliness of the environment, animal protection, the ability to hunt for culturally important food species undisturbed by visitors;
- development – tourism and employment opportunities for the community;
- culture – providing advice on Inuit perspectives of park management, providing knowledge about the historical importance of the area to Inuit (e.g. the trails), protection of heritage; and
- engagement – local engagement to support community development and safety.

These aspirations match those espoused in the IIBA. There are also interconnected, as evidenced in one community member's overall vision for the park:

> all of these Nunavut parks are unique in terms of cultural, traditional and Inuit way of survival; therefore, I would integrate and implement our culture in the national parks…I would build *qammaqs* [semi-permanent tents], I would have local people in the park teaching tourists, visitors from other parts of the world, how we used to live, how we used to survive… it's mainly our heritage, the Inuit way of life that should be taught in the park.

The extent that existing governance groups facilitated the attainment of these aspirations was discussed at length. All interview and workshop participants stressed the importance of community involvement in park management. Their involvement in governance groups had multiple positive outcomes related directly to key aspirations. These included:

- knowledge – increased influence of knowledge on park management;
- well-being – improvements in trail condition facilitating travel between the two communities, updates on wildlife from park staff (especially on the movements of polar bears which are a safety concern), the ability to work together with other community organisations (e.g. the hamlet) to facilitate better visitor experiences, the protection of Caribou breeding grounds from mining interests;
- development – jobs that recognise the skill of local people;
- culture – opportunity to recognise and protect the history of the area and the knowledge elders have of it, including sacred campsite rings; and
- engagement – improvements in community engagement after the IIBA.

In all cases, those involved in the Joint Park Management Committee felt their views were heard. Despite the significance of these accomplishments, to be truly adaptive in both governance and management, co-management must be recognised as a process of incremental learning and problem solving (Turner and Berkes 2006; Berkes 2009; Nkhata and Breen 2010) with formal reflection on potential areas for improvement. During discussions of key points raised in interviews, community members were in agreement that they were still working out how best to work together to achieve their vision for Nunavut, for their communities, and for the park. Despite clear purposes of governance groups (statutory and informal) for park management, members' views on their role, and therefore their assessment of their experiences, differed.

The JPMC was seen as a long-term advisory structure. As described by one staff member, the JPMC

> help with decision making and direction of decision-making processes involving all aspects of Auyuittuq National Park – not in the day to day operations but in the higher level overall things that go on in the park, issues that come up, and helping to shape the direction of the park management...they are involved at a much higher level in the overall direction of the park, and they work together with Parks Canada to meet those goals in a consensus based system and decision-making process.

Research participants agreed with this depiction, stating that the Joint Park Management Committee purpose was to take a longer term view, and influence management, with the Parks Planning Team reporting to the JPMC. As one JPMC member stated "things need to be dealt with by Inuit, approved by Inuit, because it is our land". However, some research participants (half of those involved in management structures) felt there could be greater community engagement to achieve community aspirations:

> The relationship between, and the responsibility of, the park planning team and the JPMC are very different and the roles are different. Because the Federal Government appoints parts of the JPMC, they have a different mandate. ...
>
> The JPMC is more involved in decision making and the development of the national park. They [the community] are supposed to have that role as consultants to relay what is supposed to be happening....there's not enough questioning of what we should be doing coming from the parks perspective.
>
> It [the JPMC] is really a government management driven committee... because they don't consult with the people, they are more or less on their own, driven by the Superintendent or the manager of the parks system...I want to make it very clear, I have nothing against the park management planning team, the staff and the management or people who work there, I think we can do better if we communicate and talk to each other and work together to develop more opportunities for the community, Nunavummiut and future generations to come.

In defence of these comments, there has been significant community input into park operations and planning. In the case of the park management plan development, eight PPT meetings, nine JPMC meetings, a joint JPMC/PPT/IKWG meeting and at least five community consultations were carried out. Further, membership of the JPMC is not based on constituency representativeness. The types of comments made most likely stem from involvement in forms of organisation somewhat new to

Inuit, who have traditionally applied consensus based decision-making in smaller family groupings rather than engaging in processes derived largely from the colonial Westminster system (Tester and Irniq 2008). Therefore, it is not surprising that some research participants articulated views that differ somewhat from the IIBA with regards to community engagement and representation, although this may change as experience with co-operative management grows.

Research participants suggested improvements in governance arrangements that would help achieve their visions for the park and its management. Comments included:

- knowledge – the potential for an IQ co-ordinator (if one does not exist), plus the need for more opportunities to share knowledge regarding the park;
- well-being – ongoing concern that visitors and some Inuit require better knowledge of the environment for safe travel on the land;
- development: increased community engagement to consider potential economic benefits that could be gained from the park;
- culture – greater teaching of park history, especially regarding place names and the rough terrain, plus the assurance of ongoing monitoring of cultural sites; and
- engagement – a lack of recent TRI-JPMC meetings, where common issues for park management could be discussed (cut in light of recent Parks Canada budget restrictions); a desire for greater local level engagement from the Auyuittuq Parks Office in JPMC and other activities to facilitate knowledge transfer and community development associated with economic benefits from the park.

Unlike the Joint Park Management Committee and the Park Planning Team, the Inuit Knowledge Working Group is an informal structure established explicitly to improve understanding of the natural environment and cultural landscape. IKWG participants described their role as advisors on Inuit history and perspectives on the park, direction of research, and having "the responsibility to provide the knowledge into the future plans of the national park". Elders' descriptions of what it means to share their knowledge matched the broad descriptions inherent within Nunavut government, as a means to reinvigorate community development:

> When IQ becomes fully utilised, all the languages, all the knowledge of Inuit is in place, the people will heal … part of that is because IQ is not in place we are still following the *Qallunaq* law, because everything has to come from outside of the territory or outside the community. Once IQ is in place, if they can understand it, everything would turn around to the point where we can fully participate in how we run our own lives. (IKWG Nov 2012).

This description of knowledge differs from western notions of knowledge, which extract meaning independently of the experiences and values that give rise to it, and result in generalizable knowledge rather than a system for understanding, interpreting and interacting with the environment (Moller 1996; Kendrick and Manseau 2008; Jacobson and Stephens 2009; Johansson and Manseau 2012).

4 Governance as a Means to Facilitate the Application of Indigenous Knowledge

Research participants commended Parks Canada's recent interest in including IQ in the development of the Park. Elders felt that the Inuit Knowledge working groups provided a mechanism where Inuit knowledge is respected, enabling them to share their knowledge, and enabling younger community members to learn about the area from elders. However, they welcomed greater staff involvement in meetings to improve understanding and transfer of their knowledge. Likewise, some local park staff felt their involvement provided a means to develop further the 'traditional' knowledge associated with their roles in park management. This would also help to ensure knowledge sharing with visitors, addressing the knowledge and well-being aspirations expressed by participants.

A second key mechanism through which knowledge is applied to management is through staff experience: six of seven staff employed at Auyuittuq at the time of writing are Inuit. Staff knowledge and experiences contributed to management through visitor briefings on both culture and travel safety, and through their own skills and capacities to effectively and safely conduct their field duties. However, some felt that their ability to apply their skills and knowledge depended on the manager at the time.

Park management involves Inuit and non-Inuit working together to represent the views of Inuit and other Canadians in Auyuittuq. However, management decisions and operations do not fit neatly into either an Inuit or a *Quallunaq* [non-Inuit] way of doing things. Some staff expressed differences in the ways they applied IQ while on the job in comparison to in their non-work activities. Examples that illustrate this point include discussions of hunting and travel on the ice:

> If you've just seen caribou, and you were on parks time and in that scenario the people in Pang are craving caribou, what would you do? …. Sure enough he's thinking about the people, so he's going to shoot them. But that's where he's already wrong. …. We kind of feel odd when we see an animal we like to eat but we can't harvest it because we are in uniform.
>
> When I was growing up, especially during the spring, we still had ice to move around with, we would sleep during the day and hunt during the nights … During the night the snow is hard, so we're using a lot less fuel to travel, plus we don't need a tent to be heated during the night because we're awake, and during the day when the snow gets soft we go back to the camp and sleep.

The Auyuittuq National Park Management Plan includes two strategies related to knowledge and culture: (1) community engagement and connection to land and marine ecosystems and Inuit culture (including visitor experience, community relationships, and marketing to visitors); and (2) gather and share knowledge and build connection to place (including the use of Inuit Knowledge, strengthening youth connection to place, strengthening Canadian connection to culture and place, and co-operation with communities on issues of common concern). The Park Management Plan is implemented over a period of 10 years, with annual park progress reports against the Plan presented to the Joint Park Management Committee. To date, they identify that strategy (1) has at best been partially met. The objectives are important as they relate directly to economic aspirations. Reports against strategy (2) indicate some components have been well addressed (e.g. Indigenous knowledge input into development of the Cultural Resource Value Statement), some have been partially met (research priority identification, and youth involvement in monitoring and evaluation), whilst others have made limited progress (monitoring programmes and Inuktitut place names). Thus, as the plan implementation continues, there is still scope for more activities related to Inuit engagement in park management as it relates to knowledge. Our discussions identified some of the barriers to this.

Sensitivities around appropriate recognition, respect for and application of IQ were evident. Some elders expressed concern that it was inappropriate for *qallunaq* (non-Inuit) to conduct and share studies that combined IQ with scientific knowledge, while staff noted that elders were somewhat cautious about them (staff) sharing the knowledge and experience of elders with visitors.

> Somebody coming from the south to get her whatever degree that she's looking for is not really providing all the knowledge or *Inuit Qaujimajatuqangit* to the young people. Also, even Inuit to date have just been listening to the *Qallunaq* who say this is how it is, although they have more knowledge than the people who come up.

Also expressed were concerns for a loss of Inuktitut language, required to appropriately convey knowledge. These concerns are well founded. Pearce et al. (2011) identified that knowledge transmission and understanding of Inuktitut is becoming limited to some facets of cultural experience, and is dependent on access to skilled teachers. For IQ to be applied in management, opportunities for intergenerational knowledge sharing and language transmission appear to be crucial.

Despite an emphasis on IQ in the management plan, IIBA and from the Territorial government, elders and community members identified a need for both types of knowledge to be applied in park management, particularly where scientific knowledge addresses new concerns such as climate change that directly affects park management.

> The elders know outside of the environment, but what is going on inside the environment, like frost and other things, that is what they don't know and would like to learn about.

5 Discussion

Our discussions indicate that co-operative management differs from the tradition of national parks being a model for the preservation of natural resources within an area. Canadian interest in the protection of Arctic ecosystems is coupled with the lived experience and history of Inuit in those places, and their need (and right) to economic and social development (but see Martin (2007) for another perspective). This does not mean these goals are incompatible. It does however mean that the loci for management is less park centred, and that successful management depends on deep levels of engagement with the community, as they come to understand how the park might contribute to their community, rather than how they might contribute to the park.

The notions of co-operative management inherent in the Nunavut Settlement are broadly evident in the co-operative management of Auyuittuq, in that there is a clear attempt to increase the voice of Nunavummiut in their own destiny. Perhaps the most challenging aspect of co-operative management in the North is that Inuit communities are new to western systems of governance (Tester and Irniq 2008), and their perceptions and expectations of it are evolving as they themselves learn to navigate them. In light of this, the combination of both formal and informal governance structures that shape the operations of Auyuittuq National Park provide opportunities for adjacent communities to explore, as we have done together during workshops, the kind of engagement and input that might be appropriate to Inuit aspirations at any point in time. Whilst we identified general support for co-operative management, a desire for broader community engagement and an implied desire for consensus based decision-making were evident. Furthermore, the extent that the JPMC can draw on the commitment for a secretariat (IIBA, section 5.1.22) is unclear. Periodic review of the IIBA means there is opportunity for arrangements to evolve as both parties (Inuit and the Federal government) reflect on their experiences. This is perhaps the most significant innovation in the new northern arrangements, allowing co-operative management to act as a process of problem solving with formalised arrangements for power-sharing, whereas such inherent flexibility is normally only evident in informal arrangements in other parts of the world (e.g. Australia).

According to some commentators (e.g. Mercer 2008), the Nunavut Land Claims Agreement provided a means for Nunavummiut to govern on their own terms, through the principles of *Inuit Qaujimajatuqangit* – the Inuit knowledge system. This was also important to community members with whom we worked. However, the notion of a *knowledge system*, as opposed to knowledge *as an object* has proven a critical challenge in environmental management globally. Differences between Indigenous and scientific ways of knowing are well documented elsewhere (Agrawal 1995; Moller et al. 2009). However, studies on how the two can be combined or applied in management to facilitate cross-cultural learning, in ways that are respectful of both knowledges, are less common (Bohensky and Maru 2011). There are several arguments for the integration of science and Indigenous knowledge, including (1) the intricate connections between global biodiversity and cultural biodiversity (Pilgrim and Pretty 2010), (2) the invaluable contributions Indigenous knowledge can make to natural resource management particularly when gaps exist in scientific understanding (Kendrick and Manseau 2008), and (3) the role that recognition of Indigenous knowledge can play in social justice (Bohensky and Maru 2011). However, studies of knowledge integration and application are in their infancy. Some scholars (e.g. Coombes et al. 2012) suggest that these Indigenous and scientific knowledges are best held in tension, arguing against the notion of 'knowledge integration' given that the relationships between Indigenous and non-Indigenous peoples are embedded in a historicism where one party dominates the terms of engagement (Escobar 1995; Briggs and Sharp 2004). Many cases of integration have resulted in de-contextualising Indigenous knowledge from the world views that ensure appropriate interpretation (Agrawal 1995; Nadasdy 1999). Others argue for greater devolution of management to Indigenous groups to facilitate cultural resilience and improve management outcomes (Jacobson and Moller 2013; McCarthy et al. 2013). Assuming equitable and meaningful knowledge integration may not be possible, the question remains of whether embedding of governance processes in a *knowledge system* can subsequently influence the application of knowledge to management.

From the perspective of Pangnirtummiut with whom we worked, the reality of Nunavut, based on a governance system of IQ, is yet to be fully realised. In the management of Auyuittuq National Park, the application of knowledge *per se* is clearly identified within the management plan, with some related objectives achieved, while others are only partially achieved. There is no doubt that *Inuit knowledge* is influencing management. However, western notions of knowledge *as an object* limit consideration of the extent to which IQ is actually applied in managing the park. The skills and experiences of staff that are critical to field operations of the park were largely derived from their experiences travelling and hunting on the land, watching and observing elders and their families – in essence those experiences are a representation of IQ. Further, the Inuit Knowledge Working Group plays a critical role in assuring a place for and respecting the knowledge, interests and concerns of elders for management of the park (Manseau et al. 2005). With greater staff involvement and continued support, these groups could play a significant ongoing role not only in park management, but also in community engagement, providing a means to foster intergenerational sharing of knowledge and strengthening use of Inuktitut. The Inuit Knowledge Working Group is thus an example of a social space that Tester and Irniq (2008) argue is required to articulate, debate and contest knowledge rather than subvert it or limit its application.

Importantly, both elders and staff identified the need for both scientific knowledge and IQ in the management of the park, noting a specific role for science to develop understanding in areas where IQ is weaker (e.g. climate change and the implications of permafrost melting). From a western governance perspective, IQ provides not only *knowledge* per se, but a robust mechanism for identifying values and processes that guide 'knowledge' gaining, interpreting and sense making: an essential part of adaptive management (O'Flaherty et al. 2008). The underpinning of management by IQ means that it becomes imbedded in a cultural process incorporating an alternative language, an experiential mode of learning that requires ongoing nourishment, and cultural principles that guide interactions between people and their landscape (Johansson and Manseau 2012). IQ may therefore prove equally relevant to the use of science to attain the broader goals of economic and social development, as well as effective management for biodiversity and landscape management (Kendrick and Manseau 2008). While embedding governance in IQ might be considered desirable at the local scale, and may provide for effective application of knowledge in management, it requires significant flexibility in organisational terms when multi-scale bureaucracies are in place (i.e. for both Nunavut Tunngavik

Incorporation and Parks Canada), and it is this tension between the local and the supra-local that becomes evident in co-operative management.

6 Conclusion

There are some lessons to be gained from interactions with Auyuittuq as an example of a new model for co-operative management of Canada's North. Firstly, we have seen that the socio-economic considerations of land rights agreements have implications for the *scope* of the exercise of managing National Parks. This focus means a shift from thinking about the park as a loci for management, to thinking about the park as a socio-ecological system, whereby the success of park management and community development are interdependent. Secondly, statutory governance arrangements that are periodically reviewed provide the flexibility for Indigenous peoples and government to learn to work together, allowing for adaptation (rather than new arrangements) as communities come to articulate which structures best meet their aspirations. Thirdly, whilst differences exist in the interpretation of knowledge, informal groups such as the Inuit Knowledge Working Group provide a means by which broader definitions, such as that of IQ, can be articulated and discussed, supporting the application of IQ as a process for learning about and interacting with the resources and heritage that give rise to unique Northern landscapes.

Acknowledgements We would like to acknowledge the contributions of all research participants, including Matthew Nauyuk, Rebecca Nookiguak, Sila Kisa and Sakiasee Sowdlooiapik, and thank Eric Joamie and Jutanie Arnakaq for translation service. A further thanks to Nancy Anilniliak, Maryse Mahy, Margaret Nowdlak, Kathy Hanson, Andrew Maher, Jane Chisholm, and Paul Ashley for background information. Parks Canada provided travel cost and field work support for this project.

References

Agrawal, A. (1995). Dismantling the divide between indigenous and scientific knowledge. *Development and Change, 26*, 413–439.

Aiatsis. (2012). *Native title and joint management. Comparison of native title and joint management arrangements in states and territories*. Retrieved March, 2012, from http://www.aiatsis.gov.au/ntru/jurisdictionaloverviews.html

Berkes, F. (2009). Evolution of co-management: Role of knowledge generation, bridging organizations and social learning. *Journal of Environmental Management, 90*, 1692–1702.

Bohensky, E. L., & Maru, Y. (2011). Indigenous knowledge, science and resilience: What have we learned from a decade of international literature on "Integration"? *Ecology and Society, 16*, 6.

Borrini-Feyerabend, G. (1996). *Collaborative management of protected areas: Tailoring the approach to the context*. Gland: IUCN.

Briggs, J., & Sharp, J. (2004). Indigenous knowledge and development: A postcolonial caution. *Third World Quarterly, 25*, 661–676.

Carlsson, L., & Berkes, F. (2005). Co-management: Concepts and methodological implications. *Journal of Environmental Management, 76*, 66–76.

Charles, C., & Trott, C. (2010). *Flexistowe to Pangnirtung: Gift of memories*. Manitoba: Lavenham Press.

Coombes, B., Johnson, J., & Howitt, R. (2012). Indigenous geographies I: Mere resource conflicts? The complexities in indigenous land and environmental claims. *Progress in Human Geography, 36*, 810–821.

Dudley, N. (2008). *IUCN guidelines for applying protected area management categories*. Gland: IUCN.

Escobar, A. (1995). *Encountering development: The making and un-making of the third world*. Princeton: Princeton University Press.

Henderson, A. (2009). Lessons for social science in the study of new polities: Nunavut at 10. *Journal of Canadian Studies, 43*, 5–21.

Jacobson, C., & Moller, H. (2013). *Two from the same cloth? Comparing the operations and outcomes of Mātaitai and Taiāpure for delivering sustainable customary fisheries*. He Kohinga Rangihau, University of Otago, New Zealand (in press).

Jacobson, C., & Stephens, A. (2009). Cross cultural approaches to environmental research and management: A response to the dualisms inherent in western science? *Journal of the Royal Society of New Zealand, 39*, 159–162.

Johansson, K., & Manseau, M. (2012). Inuit safety culture and its relevance to management. *Society and Natural Resources, 25*, 176–190.

Kendrick, A., & Manseau, M. (2008). Representing traditional knowledge: Resource management and Inuit knowledge of barren-ground caribou. *Society and Natural Resources, 21*, 404–418.

Langdon, S., Prosper, R., & Gagnon, N. (2010). Two paths one direction: Parks Canada and Aboriginal peoples working together. *The George Wright Forum, 27*, 222–233.

Legare, A. (1996). The process leading to a land claims agreement and its implementation: The case of the Nunavut Land Claims Settlement. *The Canadian Journal of Native Studies, 16*, 139–163.

Manseau, M., Ayles, B., & Parlee, B. (2005). Traditional ecological knowledge in resource management. In F. Berkes, R. Huebert, H. Fast, M. Manseau, & A. Diduck (Eds.), *Breaking ice: Renewable resource and ocean management in the Canadian North* (pp. 143–166). Calgary: University of Calgary Press.

Martin, T. (2007). Co-management of a national park: The Wapusk National Park's experience. *Canadian Studies, 61*, 139–148.

McCarthy, A., Hepburn, C., Scott, N., Schweikert, K., Turner, R., & Moller, H. (2013). Local people see and care most? Sever depletion of inshore fisheries and its consequences for Maori communities in New Zealand [Electronic version]. *Aquatic Conservation: Marine and Freshwater Ecosystems*. doi:10.1002/aqc.2378.

McCay, B. J., & Jentoft, S. (1996). From the bottom up: Participatory issues in fisheries management. *Society and Natural Resources, 93*, 237–250.

Mercer, S. A. (2008). *Claiming Nunavut 1971–1999*. Victoria: Trafford Publishing.

Moller, H. (1996). Customary us of Indigenous wildlife – Towards a bicultural approach to conserving New Zealand's biodiversity. In B. McFadgen & B. Simpson (Eds.), *Papers from a seminar series on biodiversity* (pp. 89–125). Wellington: Department of Conservation.

Moller, H., Lyver, P., Bragg, C., Newman, J., Clucas, R., Fletcher, D., Kitson, J., McKechnie, S., Scott, D., & Rakiura Titi Islands Administering Body. (2009). Guidelines for cross-cultural participatory action research partnerships: A case study of customary seabird harvest in New Zealand. *New Zealand Journal of Zoology, 36*, 211–241.

Murray, D. (2010). Canada's northern national parks: Unaugmented landscapes, unforgettable experiences, wilderness and homeland. *The George Wright Forum, 27*, 202–212.

Nadasdy, P. (1999). The politics of TEK: Power and the "integration" of knowledge. *Arctic Anthropology, 36*, 1–18.

Nkhata, B. A., & Breen, C. (2010). A framework for exploring integrated learning systems for the governance and management of public protected areas. *Environmental Management, 45*, 403–413.

O'Flaherty, M., Davidson-Hunt, I. J., & Manseau, M. (2008). Indigenous knowledge and values in planning for sustainable forestry: Pikangikum First Nation and the Whitefeather forest initiative [electronic version]. *Ecology and Society, 13*, 6.

Parks Canada. (1982). *Auyuittuq National Park reserve interim management guidelines*. Gatineau: Parks Canada.

Parks Canada. (2010). *Auyuittuq National Park of Canada management plan*. Gatineau: Parks Canada.

Parks Canada, & Qikiqtani Inuit Association. (1999). Inuit impact and benefit agreement for Auyuittuq, Quttinirpaaq and Sirmilik National Parks.

Pearce, T., Wright, H., Notaina, R., Kudlak, A., Smit, B., Ford, J. D., & Furgal, C. (2011). Transmission of environmental knowledge and land skills among Inuit men in Ulukhaktok, Northwest Territories, Canada. *Human Ecology, 39*, 271–288.

Pilgrim, S., & Pretty, J. N. (2010). *Nature and culture: Rebuilding lost connections*. London: Earthscane.

Plummer, R., & Fitzgibbon, J. (2004). Co-management of natural resources: A proposed framework. *Environmental Management, 33*, 876–885.

Statistics Canada. (2012). Pangnirtung, Nunavut (Code 1293) and Nunavut (Code 62) (table). Census Profile. 2011 Census. Statistics Canada Catalogue no. 98-316-XWE. Ottawa. Retrieved October 24, 2012, from http://www12.statcan.gc.ca/census-recensement/2011/dp-pd/prof/index.cfm?Lang=E

Tester, F. J., & Irniq, P. (2008). Inuit Quajimajatuqangit: Social history, politics and the practice of resistance. *Arctic, 61*, 48–61.

Tester, F. J., & Kulchyski, P. (1994). *Tammarnit (Mistakes): Inuit relocation in the Eastern Arctic 1939-63*. Vancouver: UBC Press.

Turner, N. J., & Berkes, F. (2006). Coming to understanding: Developing conservation through incremental learning in the Pacific Northwest. *Human Ecology, 34*, 495–513.

Usher, P. J. (2000). Traditional ecological knowledge in environmental assessment and management. *Arctic, 53*, 183–193.

Weitzner, V., & Manseau, M. (2001). Taking the pulse of collaborative management in Canada's national parks and national park reserves: Voices from the field. In D. Hamon (Ed.), *Crossing boundaries in park management* (pp. 253–259). Michigan: The George Wright Society.

Wenzel, G., Weighs, F., & Rigby, G. (2008). *The use of traditional ecological knowledge and Inuit Qaujimajatuqangit in the management of wildlife of Nunavut: A critical review*. Iqaluit: Nunavut Wildlife Management Board.

Regional Governance and Indigenous Rights in Norway: The Finnmark Estate Case

Eva Josefsen, Siri U. Søreng, and Per Selle

Abstract In this chapter, we investigate how a regional political and administrative co-management reform in Norway, established to arrange for dialogue, cooperation and interaction between the indigenous Sami and the state was imprinted by conflict and resistance from a large part of the local population. The Finnmark Estate (FeFo) was established in 2006 in which implementation of indigenous rights in the management of land and natural resources are one of the core characteristics. The institution operates in an area where the population is ethnically mixed and where the indigenous people are a minority. Drawing on a governance perspective which recognizes the multitude of stakeholders, concerns and interests, this chapter highlights the complex interplay between legally adopted indigenous rights, which are formalized in the regional governing system, and the informal norms and barriers for change institutionalized in the surrounding society. We look into the public's attitude towards the new regional governance body that shall secure indigenous rights along with other obligations. Has the opinion changed its positions since 2006, when there was a loud opposition against the establishment of FeFo? And more importantly, under what conditions could a co-management structure which includes regional government and indigenous representatives gain public legitimacy? The chapter builds on document studies and a unique survey which explores the population's attitude toward the Finnmark Estate.

This article was written as a part of the project *Finnmarkslandskap* (*Finnmark landscape*) at Norut (Northern Research Institute) and was financed by the Regional Research Fund of North Norway.

E. Josefsen PhD (✉)
The Arctic University of Norway, Tromsø, Norway
e-mail: eva.josefsen@uit.no

S.U. Søreng
Division of Reindeer Management, The Norwegian Agriculture Agency, Alta, Norway
e-mail: siri.ulfsdatter.soreng@landbruksdirektoratet.no

P. Selle
Norut (Northern Research Institute), Alta, Norway

University of Bergen, Bergen, Norway
e-mail: per.selle@uib.no

T.M. Herrmann, T. Martin (eds.), *Indigenous Peoples' Governance of Land and Protected Territories in the Arctic*, DOI 10.1007/978-3-319-25035-9_2

Keywords Sami • Indigenous land rights • Governance • Finnmark Estate • Public trust • Legitimacy

1 Introduction and Background

Investigations of indigenous rights and consultations aiming at legal changes and implementation of Sami rights have been prioritized by the Norwegian Sami in their political struggle. The point of departure for such a strategy has been anchored in the fact that Sami do not live in territorial enclaves, but are in most local communities living together with the majority population. When the boundaries of living areas are unclear there is probably a greater risk of losing by bringing indigenous rights claims into the court system. Furthermore, the potential for learning about Sami perspectives within the Norwegian decision system is more likely emphasizing negotiations and dialogue as compared to legal claims. The Sami political system has for quite some time insisted on self-determination and gradually the Sámediggi[1] (most often translated as Sami Parliament in English) has increased its decision making power. When it comes to the question of land possession, self-determination does not mean non-interference from the majority, but is more a question of relational autonomy and non-domination (Young 2004; Broderstad 2008). In general increased Sami influence to some extent challenge the liberal democracy idea of one man – one vote since it also emphasizes collective rights.

This chapter explores the criteria for success and failure of co-management bodies established to implement indigenous land rights. The empirical setting is Finnmark,[2] the northernmost county in Norway, which is a core area of the Sami. From about 1850 the Sami were exposed to extensive 'norwegianizing' (Minde 1999; Jentoft et al. 2003; Minde 2005; Minde et al. 2008). Intermarriages between Norwegians, Finnish immigrants and Sami, in addition to the state-driven assimilation policy, have over the last two centuries to a certain degree blurred ethnic boundaries. In 2006, a new co-management institution, The Finnmark Estate agency (FeFo), was implemented to some extent also to compensate for historical injustice.

[1] The Sámediggi is a democratically elected body comprised of 39 representatives elected from 7 districts every four years. Only those listed in the Sámediggi's electoral roll have the right to vote. The registration is voluntary, signing on two criteria. First, the voters have to declare that they regard themselves Sami. Second, the voter have to confer that one self, one of the parents, grandparents or great grandparents use or used Sami as home language. One can also register if one of the parents has been registered in the electoral roll. The electoral rolls are a voter registration, not a register of the total Sami population. Established in 1989, the Sámediggi is the main political tool for strengthening the Sami's political, social and cultural position. The central government has transferred authority to the Sámediggi in some areas, primarily those concerning preservation of Sami cultural heritage, education, language and culture. The Sámediggi is a mandatory body to be consulted on matters of special concern to the Sami population (www.samediggi.no).

[2] Finnmark is one of 19 counties in Norway. It is the biggest county in Norway (48,649 km^2), and has a scattered population of about 74,000 inhabitants.

The FeFo case is the only example of Sami co-management in managing traditional areas in Norway.[3]

The management of land and resources in Finnmark is rare compared to other indigenous regions in the circumpolar north. The long term colonization of Finnmark was both economically and politically motivated (NOU 1994:2, chap. 2; NOU 2008:5, chap. 5; Bull 2011). Firstly, the region is very rich on natural resources, and secondly, colonization was instrumental in drawing the Norwegian borders in the north. The availability of jobs in the fisheries attracted people from other parts of Norway, as well as Finnish immigrants, especially in the eighteenth and nineteenth centuries (Niemi 1992). It is due to this particular history that the Sami does not live separate from the Norwegian majority population, but lives in more or less ethnically mixed local communities. Especially in the coastal communities the population's ethnicity are blurred. Politically, the Sami in Norway today participate in the local, regional and national political life to the same extent as the majority population (Selle and Strømsnes 2010). This has not always been the case, and socially, even today, the Sami still experience forms of discrimination (Hansen et al. 2008). This lack of equal status can be traced far back in time to when the policy of norwegianization started, actively suppressing Sami culture and language. The Sami struggle has thus been to enforce indigenous rights and Sami culture into the more general public policy. And importantly, since there are no Sami management territories like those in other indigenous regions in which indigenous people live more or less alone, there have been no real claims of territorial autonomy arrangements among the Norwegian Sami. Therefore, the concept of "protected areas", which is so important for many indigenous people, is not describing the structural position of the Sami. That is our point of departure.

In Finnmark, most of the land was considered crown land until the passing of the Finnmark Act in 2005. The Act is a result of a long Sami struggle to ensure rights to land and natural resources in Sami core areas. In 1984 the state put down a committee broadly composed by both Sami, regional and state representatives to investigate Sami land rights in Finnmark. Twenty years later the Finnmark Act was passed and FeFo was constructed through negotiations. Here, the Norwegian Parliament's standing committee of justice consulted the Sámediggi and the Finnmark County Council. The process resulted in the state turning all the previous considered crown land over to the inhabitants of Finnmark, approximately 95 % of the county land area or 46,000 km^2. As a result we saw the rise of a new and important co-management institution, *FeFo*. This is in fact the largest decentralizing processes of land ownership ever taken place in Norway, ending up in a broad political consensus about how to implement indigenous land rights in Finnmark. Prior to the passing of the Act, however, the public discussion in Finnmark on indigenous rights in general and the Finnmark Act in particular, were intense and conflict oriented.

[3]There are however newly established local boards managing each national park in Norway, as a delegated state task. Sami representation is one out of several sectional interest groups. The first board was appointed in 2010, and since then the Sámediggis has appointed from one to four board members in all together 41 Boards. The number of board members varies, amongst other depending on how many municipalities that are represented in the board (e-mail information from the Sámediggi in Norway on 21st of March 2013).

This chapter will examine some of the criteria for success and failure for institutions such as FeFo. The FeFo case is a prime example on how indigenous people may be included in the management of land and natural resources in areas in which the population is mixed and the indigenous people are in minority. At the same time it is also a good example of a much contested institution in which large parts of the public were critical concerning the question of implementing Sami rights. As FeFo has existed for more than 7 years, and one may assume that its management structure somehow has been set, we ask what are the public's opinions about FeFo? To what extent do the people support the management institution? And, just as importantly, under what conditions could arrangements of this type gain (increased) public support?

The next section starts by presenting the study's methodology and analytical perspective. Then we turn to the political processes which resulted in the passing of the Finnmark Act and the establishment of FeFo before we look into the formal organization of FeFo and its activity. Thereafter we analyze what the survey data tells about the legitimacy of FeFo among those the institutions is meant to serve; i.e. the inhabitants of Finnmark. The concluding section highlights and discusses the main findings and possible long term consequences of FeFo's weak public support and, furthermore, discuss how institutions like FeFo, may improve its legitimacy.

2 Method

The data behind this chapter are twofold. First, we build on secondary data; a study of the consultation process which took place in the Norwegian Parliament Standing Justice committee in the period of 2003–2005 (Josefsen 2008), and an evaluation of the Finnmark Estate in the period of 2006–2009 (Nygaard and Josefsen 2010). Then we look into the attitudes of the inhabitants living in the area. Here we use a survey carried out (by Norut Alta) in Finnmark County during the autumn of 2012. The questionnaire consisted of 48 questions, mainly closed questions, each covering different aspect of FeFo structure and activity. Out of a population of a total of 74,000, a random sample of 3000 persons over the age of 18 were selected. Answers were collected from 953 persons; generating a response rate of 33 %.

In our survey, 19 % of the respondent stated their origin for being Sami or of mixed origin, and 18 % of the respondents (N = 162) are registered in the Sámediggi's electoral roll.[4] In 2011 the electoral roll contained of 7 388 Sami in Finnmark. In order to be able to register, one has to identify as Sami, speak Sami at home, or at the very least have a relative who spoke the language at home, be they a parent, grandparent or great-grand parent. As illustrated in the figure below (Fig. 1), those who are registered in the Sámediggi's electoral roll in the survey data, is somewhat overrepresented compared to the part of the Finnmark population who are registered in the roll.

[4]According to Selle and Strømsnes 2010, those registered are in general more political active and positive towards the Sámediggi as compared to non-registered Sami.

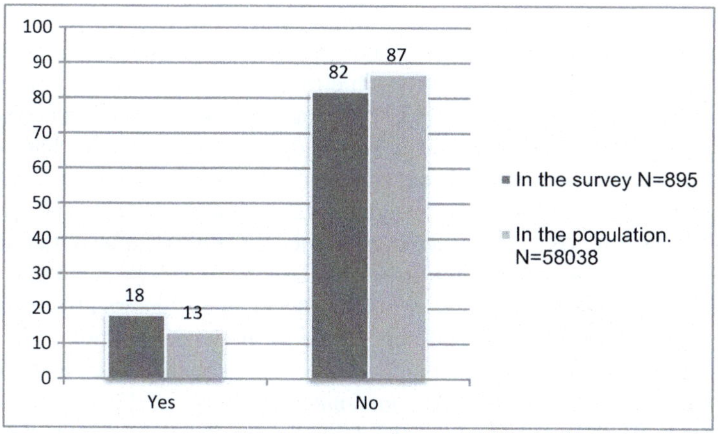

Fig. 1 People registered in the Sámediggi electoral roll, in the survey and in the general population (Percent)

Due to practical reasons will those who are registered in the Sámediggi electoral roll be referred to as Sami in our analysis, and those who are not registered in the electoral roll will be referred to as non-Sami.

3 Analytical Framework

Regional governance institutions such as FeFo forms a complex system that, drawing on Kooiman et al. (2005) and Jentoft (2007), may be described as a relationship between a 'governing system' and a 'system to be governed'. While the former is a social system made up of management institutions with their legal, administrative and knowledge systems, the latter is partly natural and partly social made up by both an ecosystem and a system of resource users and stakeholders and their knowledge and legal systems. In this chapter we are concerned with both systems, and especially the relationships between them, as we not only focus on FeFo as a formal organization, but also see it in relation to the wider institutional and societal context. That is an important point since any governance systems are infused with different sets of values, norms and principles (Kooiman and Jentoft 2009). In other words, we are here mainly concerned with how the 'system to be governed', i.e. the public, are relating or responding to the 'governing system', i.e. FeFo.

FeFo possesses almost all of the land in Finnmark and is thus the county's biggest property owner, and has an impact on important institutional stakeholders and of course the public in general. Following Hirschman (1970), people can choose loyalty, exit or voice when new management regimes are implemented (Rose 1994: Jentoft 2000: Søreng 2007, 2008). People can be loyal, which means they adapt to and follow new regulations. Or they may violate them, i.e. exit or drop out. But sometimes people also voice their opinions. This could take different forms, for instance in this case through a petition which was organized in Finnmark against the passing of the Finnmark Act in 2005, or through letters published in newspapers.

Today there is little research-based knowledge of what are- and what explain the public's attitude towards FeFo. For many, their way of thinking have most certainly links back to the state driven norwegianization policy, which lastet up until the Second World War. This to a great extent formed people's attitude to the Sami's and also strongly influenced Sami identity itself (Eidheim 1971, Otnes 2006; Eythórsson 2003). In this chapter the focus is not really on these historical processes even though we recognize their crucial importance for how people think even today. To get insight into the public opinion on FeFo, we will primarily investigate conditions concerning the *initiating* and the *implementing* phase of FeFo, and moreover, emphasize both institutional and societal conditions. People's attitude towards new management regimes could be mainly due to factors concerning the (implementation of the) formal organization of the institution and its policy. But it could as well be important contextual conditions concerning the institutions' origin, i.e. the initial phase.

Lessons from resource management studies shows that conditions concerning the pre-implementation phase may to a large extent explain public attitude towards new management institutions long thereafter (Chuenpagdee, Pascual-Fernández, Szeliánszky, Alegret, Fraga and Jentoft 2013). The opinion that stakeholders have in the initial phase of a new management institution, may remain even in the operational phase. Thus, *the images* people have about what an institution is and should be, determine their opinion on it (Jentoft et al. 2012). This means of course that a governance institution such as FeFo is not implemented in a social and political vacuum, and furthermore, that the images people have of FeFo as a management institution, being positive, negative or even indifferent, are important for its long term success. If the public, for instance, have an image of FeFo as an agency that will be to their benefit, the possibilities for support is high. If, on the other hand, their image of FeFo is that it would be to their disadvantage, it will not gain support and legitimacy. The basic understanding here is that the images the public held of FeFo in its early phase, remains also in the operational phase. This means that an alienated and divided community is, assumingly, in most cases not a good breeding ground for long term success and public trust (Chuenpagdee et al. 2013).

By focusing on both the initial and operational phase of FeFo one can put light on and perhaps even to some extent explain how new governance institutions such as FeFo, having one of its core characteristics to implement indigenous land rights, could gain public trust. Selznick (2003) holds that social support is crucial for a well-ordered legal system, as FeFo could be regarded as. Social support is thus important for long term institutional survival. His reasoning is that "the more integrated law is with other institutions, and with what people can accept as sensible, the easier it is to make the system work, and to deliver justice as well as law." (Selznick 2003: 178; see also Habermas 1996). This is highly relevant for the new co-management structure in Finnmark. We are talking about processes which do not take place in a social vacuum, but are engraved in contexts of institutional practices that, following Scott (1995), are infused by different regulative, cultural and cognitive patterns. If the values, norms and principles the public hold towards indigenous rights depart from those represented by management institutions, here FeFo, it could lead to a breakdown in public support and lack of legitimacy for its policy (Søreng 2006, 2013).

4 Implementing Indigenous Rights

4.1 The Passing of the Finnmark Act

The starting point of the Finnmark Act and FeFo can be traced back to the conflict in the 1980s regarding the government proposal to dam the Alta River and build a hydroelectric power station in core Sami settlement areas. This raised a political storm and resulted in active civil disobedience.[5] The conflict was from the beginning both a Sami rights issue and an environmental issue, and mobilized both Sami as well as non-Sami. The Sami mobilization challenged the state land ownership, and also put the Sami issue into a larger picture of Sami language, culture and livelihood. The controversy made the Sami question almost overnight a matter of intense debate also at the national level, and resulted in several arrangement to secure Sami rights, as among other: the establishment of the Sámediggi in 1989; the ratification of the ILO Convention No. 169 concerning Indigenous and Tribal Peoples in Independent Countries in 1990; the amendment of the Sami Paragraph in the constitution in 1988; and consequently the Finnmark Act in 2005.

After more than twenty years of public investigation of Sami land right issues in Finnmark, the Norwegian Parliament's Standing Committee on Justice handled the Finnmark Act case from 2003 to 2005, ending with the final proposal passed by the Parliament. The legislative process in the committee turned out to be an event without precedence because both the Sámediggi and the Finnmark County Council participated through four consultation meetings. The Norwegian Parliament had never before allowed "outsiders" to interfere at this final stage of a legislative process, a process that had a considerable effect on the original proposal and in the end producing legislation which brought Sami land rights into alignment with international law on the rights of Indigenous peoples (Josefsen 2008, 2011). The legislative process leading up to the Finnmark Act showed that it was the Sámediggi and not the Finnmark County Council that was producing new perspectives and demands on land ownership and management. This innovative cooperation between the justice committee, the Sámediggi, and the Finnmark County Council which was within the formal framework of the ILO-k 169, was strongly based on knowledge sharing and a long term trust building between the partners (Josefsen 2011). Especially the

[5] For more information see Svensson 2002; Paine 1982.

Sámediggi had prioritized establishing such a foundation for dialogue with the justice committee. In the final period, the Sámediggi got in a position where they negotiated the last changes in the Act proposal with the committee (Josefsen 2008). The Finnmark Act is thus in its final version grown out of the question of indigenous land rights. The purpose of the Finnmark Act is:

> (..) to facilitate the management of land and natural resources in Finnmark in a balanced and ecologically sustainable manner for the benefit of the residents of the county and particularly as a basis for Sami culture, reindeer husbandry, use of non-cultivated areas, commercial activity and social life (Preamble to the Finnmark Act).[6]

With the passing of the Finnmark Act, the ownership of former state land was transferred to FeFo. The Act is however ethnical neutral when it comes to which individuals have the right to use land managed by the FeFo (Josefsen 2007).

4.2 The new Governance Institution: Tasks, Policy and Formal Organization

FeFo consists of an administration of approximately 35 employees. The administration was inherited from the prior State land regime in Finnmark that was abolished when the Finnmark Act was passed. The executive board of FeFo consists of six members, three appointed by the Sámediggi and three appointed by the Finnmark County Council. It is important to note that the ownership of the prior state land was not transferred to the Samis, but to FeFo that manages this property on behalf of the whole population of Finnmark. FeFo is mainly a private landowner which has the same relationship to the public authorities as other landowners.

FeFo is managing both land-use and all the natural resources on land.[7] Residents of the county are entitled to exploit the natural resources on FeFo land, through activities such as hunting, fishing or cloudberry picking. The Finnmark Act gives the local population, without treating inhabitants differently on the basis of ethnicity, greater rights to exploit renewable resources in the county than has previously been the case. The general public outside the county still enjoy right of access for purposes such as sports fishing in inland waters, small-game hunting and trapping (§21–27 in the Finnmark Act). Furthermore, the Finnmark Act acknowledges land rights acquired by the Sami and others through customary or ancestral use. These rights still await judicial clarification as to whether they are user or property rights.

---

[6] The Act also states that "the Act shall apply with the limitations that follow from ILO Convention No. 169 concerning Indigenous and Tribal Peoples in Independent Countries" (section 3).

[7] When it comes to minerals it is according to the Norwegian Mineral Act divided into two groups; the states minerals (minerals with own weight over 5 kg pr. liter and where iron, copper, sink, silver, gold and lead are the most common) and the land owners' minerals (all minerals that do not belong to the state, for example quartz, quartzite, nepheline, diamonds and natural stone, together with sand and gravel). The state minerals are claimable by searchers, while FeFo have possession of all other minerals.

A separate Commission is set up for this purpose. A Land Title Court is also created to handle disputes over ownership or use that may emerge in the wake of the Commission's investigation. The Sámediggi has the right to issue guidelines for any changes to the use of land (section 4 in the Finnmark Act) and FeFo is obligated to evaluate what impact important changes in land use will have on Sami culture, reindeer husbandry, use of non-cultivated areas, commercial activity and social life more in general (section 10 in the Finnmark Act).

As mentioned, FeFo is run by a board, appointed by the County Council and the Sámediggi. However, since FeFo is a foundation, neither the Sámediggi nor the Finnmark County Council can instruct any of the board members. The FeFo board is not responsible for any electorate and the County Council and the Sámediggi have no other formal role than to point out board members. Since the board, in its first years, to a large extent has emphasized consensus in its decision making processes, and not politicizing along ethnical or political party lines, the only way to influence board decisions have been through dialogue. One of the major challenges in the first years of FeFo was the establishing of norms for communication between the County Council, the Sámediggi and the FeFo board members (Nygaard and Josefsen 2010). Especially the Finnmark County Council and its elected politicians had a hard time accepting that they did not have any authority in terms of giving directions to the County Council appointed FeFo board members, and disagreements were played out in the open through the media. The interaction between the three institutions is now organized into half year meetings between the FeFo board, the Sámediggi and the Finnmark County Council.

5 Indigenous Rights and Public Opinion

5.1 The General View and Knowledge About FeFo

It was not only conflicts among the cooperating institutions that soon became visible in the media. In the media there has also been forwarded a strong and consisting resistance against the Act among parts of Finnmark's population (Eira 2013). This opposition had a peak in the spring of 2005 when the Norwegian Parliament passed the bill, an opposition that stood in contrast to the positive co-operation climate seen in the political consultation process of 2003–2005. Sami rights were contested by the public. The resistance materialized through a petition against the Act which gathered more than 10 000 signatures in Finnmark of a total population of 74 000. It also led to the foundation of a new organization in Finnmark (EDL) protesting against legalized Sami rights and the indigenous status of the Sami. Public voices challenged both the indigenous status of the Sami, and also the Sámediggi's power concerning board members, emphasizing the fact that the Sami is a minority within the county. In spite of this resistance, most of the popularly elected representatives in the Finnmark County Council and the Sámediggi were satisfied with the bill and supported it

(Ravna 2005:225). This cleavage between a supporting political elite and a skeptic public is seemingly still present, and distinctly voiced through the media.

In the autumn of 2012, a survey explored the inhabitants' opinion, knowledge of, and experiences with FeFo. This is the first and so far only research based initiative to measure people's attitude towards this new governance system. The respondents reflect the Finnmark population, consisting of both Norwegian, Sami, Kven,[8] and newer immigrants. 18 % of the respondents (N=162) are registered in the Sámediggi's electoral roll.

Table 1 shows that in Finnmark, there are more people who are dissatisfied (28 %) than satisfied (20 %) with the work FeFo does today. Many are however neither more nor less satisfied with FeFo's work (34 %), and one fifth has not made up their mind (18 %).

It turns out that Sami respondents tend to be more satisfied with FeFo's work (36 %) than the rest of the respondents. There are also fewer among the Sami who are dissatisfied (20 %).

The picture is diverse when looking at the public's level of knowledge of what FeFo is actually doing. Table 2 shows that one third describes their knowledge of FeFo's activities as limited. About the same numbers describe their knowledge being neither more nor less, while those who say they are well acquainted with FeFo is 29 %.

There are relatively more Sami who say their knowledge about FeFo's activity is rather good (40 %), and there are fewer Sami that consider it to be limited (22 %).

In table 3 we look at the respondents' knowledge of FeFo's activity and their support of FeFo's work. Among those who are dissatisfied with FeFo's work, 42 % describe their knowledge as good, compared to 32 % among those who are satisfied with FeFo's work. However, only 10 % of those who are satisfied with FeFo's work describe their knowledge of FeFo as limited, compared to those who are discontent with FeFo, where 18 % say they have limited knowledge. Among those who have not made up their mind concerning whether they are satisfied with FeFo's work, there are more people who have limited knowledge of FeFo's activity (31 %) than those who says they have good such knowledge (25 %). Almost half of those who were neither/nor pleased with FeFo's work also state that they have neither/nor knowledge about the management institution. In general then, we cannot conclude from these numbers that the more one knows about FeFo's activities the stronger

Table 1 I am satisfied with the work of FeFo (In percent; N=953)

Completely or almost completely agree	20
Neither/nor	34
Completely or almost completely disagree	28
Do not know	18
Total	100

[8]The Kven is a minority group origin from Finland. They have immigrated over the last centuries.

Table 2 The Finnmark people's knowledge of FeFo's activities (In percent; N=953)

Good knowledge of FeFo's activity	29
Neither/nor	34
Limited knowledge of FeFo's activity	34
Do not know	3
Total	100

Table 3 The Finnmark people's knowledge of FeFo's activities and view's on FeFo's work (In percent; N=766)

		The Finnmark people's knowledge of FeFo's activities			
		Good	Neither/nor	Bad	Do not know
Finnmark people's view on FeFo's work	Satisfied	32	18	10	5
	Neither/nor	25	49	31	5
	Dissatisfied	42	26	18	19
	Do not know	1	7	41	71
	Total	100	100	100	100
		N=231	N=265	N=108	N=21

one will support the institution. In other words, this does not give us sufficient grounds to conclude that people's self-reported knowledge about FeFo's day-to-day management activities have an impact on how they evaluate FeFo's work.

From the table above one get the impression that people who are well acquainted with FeFo's activities also are among those who to a greater extent have made up their mind whether they are pleased or not with FeFo's work more in general. The relatively high percentage of respondents who are dissatisfied with FeFo's work (28 %), (Table 1), has to be explained in other ways than lack of knowledge. An interesting question is, therefore what is the information source for people's knowledge of FeFo.

The survey shows that people's knowledge about FeFo comes from various formal and informal channels. Figure 1 shows that most people get their knowledge through media such as newspapers, radio and TV (77 %), followed by family members and/or friends (28 %). Some have gained knowledge from formal sources, such as FeFo's written information (23 %), FeFo's homepage (15 %) or by themselves consulting FeFo directly (9 %). There are also some who have gained their knowledge through information from political institutions such as County Council (10 %) and the Sami Parliament (3 %). There are no real differences between the registered Sami and the rest of the respondents on these matters (Fig. 2).

5.2 *What Does our Data Tell About the Legitimacy of FeFo*

According to the survey, the population is split in their views on whether an institution such as FeFo is really needed in Finnmark. Table 4 shows that a third believes it is needed, while more than an third says it is not. Almost one fifth answers neither/

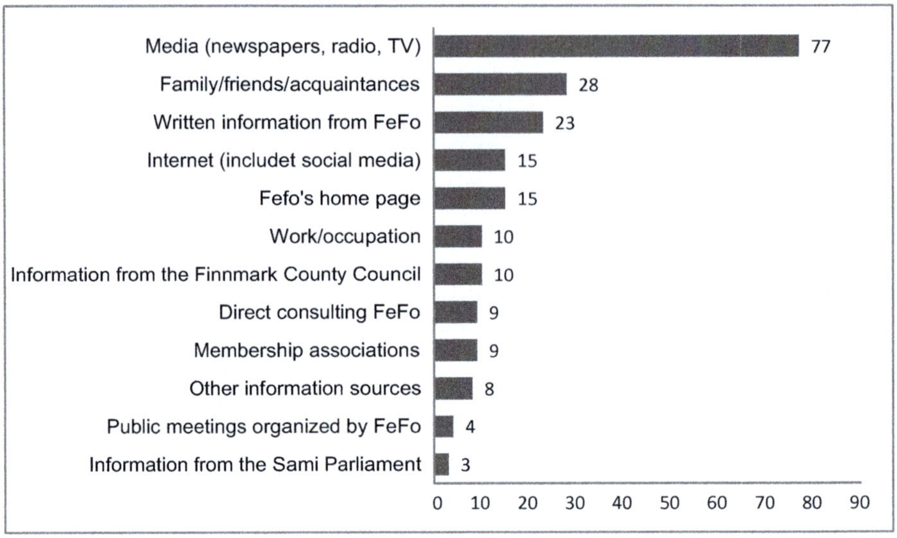

Fig. 2 Information sources concerning Finnmark people's knowledge about FeFo (In percent; N=953)

Table 4 Sami's and non-Sami's view on the need of FeFo (In percent; N=720)

	Total	Registered Sami	Non-Sami
Need FeFo	31	38	30
Neither/nor	19	23	18
No need of FeFo	35	23	38
Do not know	15	16	14
Total	100	100	100
	N=720	N=137	N=582

nor, while 15 % answer don't know. The table also shows that there are more registered Sami (38 %) than non-Sami (30 %) who say there is a need for a management institution like FeFo. Likewise there are more non-Sami (38 %) than Sami (23 %) who believe there is no need of FeFo.

In the survey (Table 5) we also asked what people thought of FeFo just before it was established, or what *image* they had of FeFo. Over half of the respondents say that they were negative to FeFo back in 2006, while only 14 % were positive. 15 % were indifferent. There is also a rather clear difference between how registered Sami and non-Sami look at FeFo back in 2006 (Table 5). The Sami (33 %) were much more positive to FeFo than the non-Sami (10 %) which also imply that the non-Sami (59 %) were more negative back in 2006 than the Sami (36 %). The Sami were much more divided in their view on FeFo in 2006 than the non-Sami, who tended to be clearly more negative.

Table 5 The public attitude towards FeFo when established in 2006 (Sami and non-Sami; In percent; N=823)

	Total	Registered Sami	non-Sami
Positive to FeFo when established in 2006	14	33	10
Negative to FeFo when established in 2006	55	36	59
Indifferent to FeFo when established in 2006	15	19	14
Do not remember	16	12	17
Total	100	100	100
	N=823	N=151	N=672

Table 6 Registered Sami's and non-Sami's view on whether FeFo should continue to exist (In percent; N=852)

	Total	Registered Sami	Non-Sami
FeFo should continue to exist	27	41	23
FeFo should close down	44	30	47
Do not know/no opinion	29	29	30
Total	100	100	100
	N=852	N=157	N=695

Furthermore, Table 6 tells that even after having existed for six years, only one fourth want to see FeFo to continue to exist (27 %). Many are not able to give a clear answer as they have not made up their mind (29 %). Most importantly, however, a large amount of Finnmark people want FeFo to wind up (44 %). This is a rather dramatic picture appearing. Sami tends to be somewhat more positive to the existence of FeFo than non-Sami (Table 6). Among the Sami, 41 % say that FeFo should continue to exist, but as much as 30 % express that it should close down. Among the non-Sami respondents, 47 % want FeFo to close down, while 23 % believe that it should continue to exist. Large percentages among the Sami and non-Sami do not know or have no opinion on this matter.

It could however seem like the public's attitude towards FeFo has changed somewhat since it was established back in 2006. Looking back at Table 5; more people were negative to FeFo (55 %) back in 2006, compared to Table 6, where 44 % want it to close up. In 2006, 33 % of the Sami were negative to FeFo – but today 41 % of the Sami want it to continue to exist. For the non-Sami, there are twice as many today that want its existence (23 %), than those who were positive to FeFo in 2006 (10 %). Even if the questions are not identical, the findings could be interpreted as if the public attitudes towards FeFo has somewhat improved. In general, though, the legitimacy of FeFo among the Finnmark population is low, and with the background of FeFo in mind, interestingly enough also the support among the Sami is limited.

6 What Are the Different Positions All About?

Before FeFo was implemented, when the content of the Finnmark Act was settled, the justice committee consultations with the Sámediggi and the Finnmark County Council had a structure that makes a governance perspective highly relevant. There were however important differences among the partners. The County Council acted mostly as a body entitled to comment, responding on questions from the justice committee. The Sámediggi was much more offensive, forwarding position reports on several topics concerning land rights in Finnmark, actively debating with the committee members, and also entering into negotiations with the committee at the end of the legal process (Josefsen 2008).

A governance perspective on the question of land rights underlines that the Sami may achieve influence without being coopted into the majority political system, even if it will always be a certain pressure in that direction. The Sámediggi insisted on not accepting the establishment of state power delegating bodies, but strongly argued for transferring land ownership to the people of Finnmark. The position itself probably reduced the risk of cooptation.[9] The significance of the ILO-k 169 in the case of the Finnmark Act is here important. This international obligation alongside with Norway's international reputation as a human rights advocate, created a specific form of interdependency. The justice committee consultations can thus be viewed as an arena connecting public policy with stakeholders, and "(..) overcoming the constrains and limitations of representative democracy and party politics" (Klijn and Skelcher 2007: 588).

The pre-implementation phase of FeFo could in many respects be described as a successful collaboration between the main institutional stakeholders resulting in the passing of the Finnmark Act. The state, the Finnmark County Council and the Sámediggi reached an agreement which resulted in the new and important governance regime in Finnmark, transferring former state land to the people of Finnmark, i.e. FeFo. However, discussions in the media about the ILO-k 169, the Finnmark Act – and the extensive petition against it – show that many voiced a strong opposition towards the new regime that was meant to emphasize indigenous rights through the regionalization of land management in Finnmark. Despite the success of the institutional collaboration, a large amount of the population, according to our 2012 survey, had a negative view of FeFo already before it came into being in 2006. Especially among the non-Sami, but interestingly enough not at all only within this group, the attitude towards FeFo was largely negative. Even though there were more Sami than non-Sami that were positive towards the establishment of FeFo in 2006, surprisingly many Sami were also negative. One could say that the new governance

[9]In other land management arrangements, for example when it comes to management of national parks in Norway, there are examples of deep disagreement between national- and local level. Such disagreement may undermine public trust in the system. Thus there have been established other co-management arrangements in which the Sami plays a role, or government arrangements, but then "in the shadow of hierarchy" (Scharpf 1994).

institution was born in a rather hostile environment meeting strong resistance from the people it was meant to serve.

There is no room in this chapter for going deeply into the reasons for this being the case, but we shall mention certain important factors. One explanation of the negative attitudes may be that the Finnmark Act was not a result of claims forwarded by the county population at large; the reform was not even supported by all Sami, as the survey points out. It was coming in from “above” and many, both Sami and non-Sami, felt the reform was enforced upon them, as a form of external coercion (Olsen 2011). Part of the explanation is also, we believe, that indigenous right is not considered just within parts of the population of Finnmark. It may even be understood as a threat towards their own land use, even though the Act itself does not separate along ethnical lines. Still, people seemed to hold the image that the Act would create important legal differences by grouping the inhabitants as “indigenous” and “non-indigenous”, and consequently believing that the new governance regime would exclude some from enjoying rights to natural resources (Benda-Beckmann 1997).[10] The strong resistance among many inhabitants in Finnmark may therefore be interpreted as a fear of weakening the democracy in land management (Olsson 2003).

To some extent people's negative image of FeFo could have been strengthen by the way the consultation process was organized. The consultations and the discussions in the justice committee were closed, and none of the participants took the responsibility to inform the public about the process. Thus, the public debate was to a large extent based on assumptions and misinformation (Olsen 2011; Fossbakk 2010). Also the way FeFo operated could strengthen peoples mistrust. In the beginning, the FeFo-board closed its meetings from the public. By doing so, the board opened the door for suspicion and public mistrust; all reflected in the media discussions. The lack of transparency probably contributed to people's negative image of FeFo in its early operational phase.

However, there are also elements of historically based attitude towards Sami rights that may be really important here. The hundred year period of norwegianization of the Sami (and also the Kven minority) have left its footprints in people's minds and made ethnic identity complicated or even a social stigma (Eidheim 1971; Kramvig 1999). This could explain why there in general seem to be a negative attitude towards the idea of indigenous rights in Finnmark and even to some extent among the Sami themselves. Out of this an interesting question arise; are the public as negative to how the new governance regime in practice *manage* the land- and resources in Finnmark, as they are to the principles of which FeFo is based on? This important question cannot be answered at this stage. More analysis has to be done, but we believe that it is an important distinction to have in mind.

The people of Finnmark's attitude towards FeFo, we believe, are also influenced by the media. The media held a conflict approach on Sami rights in the early phase; i.e. when the substance of the new Finnmark Act was debated. Newspapers were

[10] Similar arguments are also put forward in the discussion about implementing Sami rights in the Finnmark fisheries (Søreng 2013).

filled up by letters discussing the Finnmark Act and FeFo, mainly published by people having a negative attitude towards legalizing Sami rights (Eira 2013). The composition of the FeFo board was an especially hot media issue, especially since the Sami may vote to both for the County Council and the Sámediggi elections, at the same time as being a minority in the County. Arguments was on the one side anchored in ideas of liberal democracy (one man-one vote), and on the other side on a view of democracy in which collective rights will secure ethnic minorities and give them cultural protection and equal status as the majority (Kymlicka 1989). Giving the historical background, including the fact that many recieved knowledge about FeFo from the media, the public's opinion on FeFo was to some extent formed by these discussions prior to the establishment of FeFo. Similar to the findings from implementation studies of Marine Protected Areas (Jentoft et al. 2012; Chuenpagdee et al. 2013), it is not only FeFo's policy and the output coming from its management activities that determine the public attitude towards the institution, but definitely also the views people held of it in the pre-establishing phase.

7 Conclusion

The institutionalizing of indigenous rights in land- and resource management is a process infused by different, and contradictory, norms, values and principles. At the institutional level, the political processes which lead to the establishment of FeFo changed from being conflict oriented to end up with an important political consensus on basic indigenous land rights. The state, the Sámediggi and the County Council succeeded in landing the Finnmark Act. This success was interestingly enough not reflected in corresponding support from the public. Even though the new regime also had its supporters, many fought with great intensity against the Finnmark Act. Resistance was voiced through a very strong petition campaign and in the media, especially the local newspapers. The Finnmark Act was in this manner well integrated with the political institutions of the regional and national level, but not in the mind of the inhabitants. It was therefore a huge gap between the agreement at the institutional level, and the different sets of norms, values and principles important for the people of the region. This was the situation when the new regional governing institution (FeFo) came into being.

The images the Finnmark inhabitants had of FeFo when established in 2006 were to some extent improved in 2012, but still the legitimacy of the institution is low. The legitimacy is not even particularly high among the Sami themselves. Furthermore, many lack core knowledge of what FeFo actually does. FeFo could to some extent be hold responsible for that situation due to scarce information coming from the institution, leaving the media as the public's number one source of information. But neither FeFo's 'founding fathers' (the state, the Sámediggi and the Finnmark County Council) have done much to improve the situation. The County Council even attacted FeFo in its early phase, while the prior land owner (the state) and the Sámediggi backed out and left FeFo to a large extent alone in defending its new role, and the Finnmark Act, in its early operational phase.

For governance institutions such as FeFo, which are established to realize indigenous rights, it is of course important to have legitimacy among the people it is meant to serve. However, trust does not rise from nothing. FeFo must, one could argue, earn the public's support. Today FeFo tends to have more legitimacy among the Sami in Finnmark than the non-Sami. Part of the reason for this difference is probably that the Sami are in general more supportive towards the basic principle in which the institution is based upon; that of implementing indigenous rights, i.e. their own rights.

As Finnmark is ethnically mixed, it will be highly problematic if FeFo in the long run lack support among different ethnic groups. However, it should again be mentioned that neither the Finnmark Act nor FeFo are ethnical discriminating when it comes to implementing their policies. If the lack of support derives from groups being suspicious of FeFo's policy is being discriminating along ethnical lines, a joint information strategy from the primary stakeholders may improve the situation.

One important lesson from this study is that new governance regimes which are implementing indigenous rights should not undervalue the significance of the long term effects of a former state driven assimilation policy towards the population at large. Such long term consequences can be perceived as an unwillingness to recognize indigenous rights in land and natural resource management, both among parts of the majority as well as among parts of the minority. Another important lesson is not to undervalue the importance of informing the public about the background for and the goals of land management reforms in a planned and targeted way, and not leaving the public to be informed mostly by conflict oriented and politicized media publicity. A broad information strategy seems to be important not only in the establishment phase of an new management institution, when people's views of new management institutions are set, but also in the operating phase. Keeping the public informed about goals, processes and outputs, may contribute to improve people's understanding and support of such an extensive reform as the passing of the Finnmark Act and the establishment of the FeFo represent, and thus, establish a sense of public "ownership" to the new management institution.

References

Benda-Beckmann, F. (1997). Citizens, strangers and indigenous people: Multiple constructions and consequences of rights, resources and people. *Law and Anthropology, 9*, 1–42.

Broderstad, E.-G. (2008). *The bridge-building role of political procedures. Indigenous rights and citizenship rights within and across the borders of the nation-state*. Tromsø: University of Tromsø.

Bull, K. S. (2011). *Kystfisket i Finnmark*. Oslo: Universitetsforlaget. [in Norwegian].

Chuenpagdee, R., Pascual-Fernández, J. J., Szeliánszky, E., ALegret, J. L., Fraga, J., & Jentoft, S. (2013). Marine protected area: Re-thinking their inception. *Marine Policy, 39*, 234–240.

Eidheim, H. (1971). *Aspects of the lappish minority situation*. Oslo: University of Oslo.
Eira, S. S. (2013). Herrer i eget hus Finnmarlsloven i media. *Norsk medietidsskrift, 20*(4), 330–346. [in Norwegian].
Eythórsson, E. (2003). The coastal sami: 'a Pariah Caste' of the Norwegian fisheries? A reflection on ethnicity and power in Norwegian resource management. In S. Jentoft, H. Minde, & R. Nilsen (Eds.), *Indigenous peoples. Resource management and global rights*. Delft: Eburon Academic Publishers.
Finnmark Act. Act of 17 June 2005 No. 85 relating to legal relations and management of land and natural resources in the county of Finnmark.
Fossbakk, O-B. (2010). *Below the public policy surface: Local reality and popular resistance against the Finnmark Act* (pp 74–77). Forum conference report 2010: Indigenous participation in policy-making: Ideals, realities and possibilities. University of Tromsø. Internett, Retrived March 14, 2013, from http://munin.uit.no/bitstream/handle/10037/2941/report.pdf?sequence=3
Habermas, J. (1996). *Between facts and norms*. Cambridge, MA: The MIT Press.
Hansen, K. L., Melhus, M., Høgmo, A., & Lund, E. (2008). Ethnic discrimination and bullying in the Sami and non-Sami populations in Norway: The SAMINOR study. *International Journal of Circumpolar Health, 67*(1), 99–115.
Hirschman, A. O. (1970). *Exit, voice, and loyalty: Responses to decline in firms, organizations, and states*. Cambridge, MA: Harvard University Press.
Jentoft, S. (2000). Legitimacy and disappointment in fisheries management. *Marine Policy, 24*, 141–148.
Jentoft, S. (2007). Limits of governability? Institutional implications for fisheries and coastal governance. *Marine Policy, 31*, 360–379.
Jentoft, S., Minde, H., & Nilsen, R. (Eds.). (2003). *Indigenous peoples. Resource management and global rights*. Delft: Eburon Academic Publishers.
Jentoft, S., Pascual-Fernandez, J. J., De la Cruz Mondino, R., Gonzalez-Ramallal, M., & Chuenpagdee, R. (2012). What stakeholders think about marine protected areas: Case studies from Spain. *Human Ecology*. doi:10.1007/s10745-012-9459-6.
Josefsen, E. (2007). *Finnmarksloven og Finnmarkseiendommen*, Periodica Plan, 2/2007. [in Norwegian].
Josefsen, E. (2008). Stat, region og urfolk – finnmarksloven og politisk makt. In H.-K. Hernes & N. Oskal (Eds.), *Finnmarksloven*. Oslo: Cappelen akademisk. [in Norwegian].
Josefsen, E. (2011). The Norwegian Sámi Parliament and Sámi political empowerment. In G. Minnerup & P. Solberg (Eds.), *First world, first nations*. Brighton: Sussex Academic Press.
Klijn, E.-H., & Skelcher, C. (2007). Democracy and governance networks: Compatible or not? *Public Administration, 85*(2), 587–608.
Kooiman, J., Bavinck, M., Jentoft, S., & Pullin, R. (Eds.) (2005). *Fish for life: Interactive governance for fisheries*. Amsterdam: Amsterdam University Press.
Kooiman, J., & Jentoft, S. (2009). Meta-Governance: values, norms and principles, and the making of hard choices. *Public Administration, 87*(4), 818–836.
Kramvig, B. (1999). I kategorienes vold. In H. Eidheim, (Ed.), *Samer og nordmenn*. Oslo: Cappelen Akademiske forlag. [in Norwegian].
Kymlicka, W. (1989). *Liberalism, community and culture*. Oxford: Claredon Press.
Minde, H. (1999). Mot rasediskriminering, for urfolksretter – To sider av same sak? Et historisk perspektiv på samiske rettigheter. In H. Eidheim, (Ed.), *Samer og nordmenn*. Oslo: Cappelens Akademiske Forlag. [in Norwegian].
Minde, H. (2005). Assimilation of the Sami – Implementation and consequences, *Gáldu Čála -Journal of indigenous rights, 3*/2005.
Minde, H., Gaski, H., & Jentoft, S. (Eds.). (2008). *Indigenous peoples: Self-determination, knowledge, indigeneity*. Delft: Eburon Academic Publishers.
Niemi, E. (Ed.) (1992). *Pomor: Nord-Norge og Nord-Russland gjennom tusen år*. Oslo: Gyldendal. [in Norwegian].

Norwegian Official Report. (1994). *NOU 1994:21 Bruk av land og vann i Finnmark i historisk perspektiv*. Oslo: Departementenes servicesenter. Informasjonsforvaltning. [In Norwegian].
Norwegian Official Report. (2008). *NOU 2008:5 Retten til å fiske i havet utenfor Finnmark*. Oslo: Departementenes servicesenter. Informasjonsforvaltning. [In Norwegian].
Nygaard, V., & Josefsen, E. (2010). *Finnmarkseiendommen under lupen*, Norut-report Nr 2010:4. [in Norwegian].
Olsen, K. (2011). Fefo, reinsdyr og andre vederstyggeligheter – Om urfolk, staten og finnmarkingene, *Norsk antropologisk tidsskrift*, *2*/2011. [in Norwegian]
Olsson, J. (2003). Democracy paradoxes in multi-level governance: Theorizing on structural fund system research. *Journal of European Public Policy, 10*(2), 283–300.
Otnes, P. (2006). Retrospect on the Sami Nation. Text, context, field, symbolic violence confessions of a bordering Actor. In V. Stordahl, (Ed.), Samisk identitet. Kontinuitet og endring. Diedut, 3, 127–155.
Paine, R. (1982). *Dam a river, damn a people?: Saami (Lapp) livelihood and the Alta/Kautokeino Hydro-Electric Project and the Norwegian Parliament* (IWGIA document; 45). Copenhagen: International Work Group for Indigenous Affairs.
Ravna, Ø. (2005). Finnmarksloven er vedtatt. Om de vesentlige endringene i loven i forhold til regjeringens forslag i Ot.prp. nr. 53 (2002–2003). Kart og Plan 4:224-228. [In Norwegian].
Rose, C. (1994). *Property and persuasion. Essays in the history, theory, and rhetoric of ownership*. Boulder: Westview Press.
Scharpf, F. W. (1994). Games real actors could play. Positive and negative coordination in embedded negotiations. *Journal of Theoretical Politics, 6*, 27–53.
Scott, W. R. (1995). *Institutions and organizations*. Thousand Oaks: Sage.
Selle, P., & Strømsnes, K. (2010). Sámi citizenship: Marginalisation or integration? *Acta Borealia, 27*(1), 66–90.
Selznick, P. (2003). "Law in Context", Revisited. *Journal of Law and Society, 30*, 177–186.
Søreng, S. U. (2006). Moral discourse in fisheries co-management: A case study of the Senja fishery, northern Norway. *Ocean and Coastal Management, 49*, 147–163.
Søreng, S. U. (2007). Fishing rights struggles in Norway: Political or legal strategies? *Journal of Legal Pluralism, 55*, 187–210.
Søreng, S. U. (2008). Fishing rights discourses in Norway: Indigenous versus non-indigenous voices. *Maritime Studies, 6*(2), 77–99.
Søreng, S. U. (2013). Legal pluralism in norwegian inshore fisheries: Differing perceptions of fishing rights in Sami Finnmark. *Maritime Studies, 12*(9), 1–21.
Svensson, T. (2002). Indigenous rights and customary law discourse. Comparing the Nisga'a and the Sámi. *Journal of Legal Pluralism, 47*, 1–35.
Young, I. M. (2004). Two concepts of self-determination. In S. May, T. Modood, & J. Squires (Eds.), *Ethnicity, nationalism, and minority rights*. Cambridge: Cambridge University Press.

Increasing Cooperation and Advancing Reconciliation in the Cooperative Management of Protected Areas in Canada's North

Thomas H.D. Nesbitt

Abstract Modern-day land claim agreements and protected area agreements create an opportunity to help reconcile the interests and worldviews of aboriginal and non-aboriginal Canadians. Among protected area agreements, three in northern Canada contribute particularly to this objective. In the Saoyú-ʔehdacho Agreement (2008), the Tuktut Nogait Agreement (1998) and the Gwaii Haanas Agreement (1993), various aboriginal authorities and the government of Canada have agreed to use the aboriginal concept and practice of consensus decision-making in their cooperative management of the vast tracts of land protected pursuant to these agreements. This paper discusses these agreements through the lenses of reconciliation, consensus decision-making and cooperative management. More specifically, it discusses the reconciliation objective underlying the recognition, in Canada's *Constitution Act, 1982*, of aboriginal and treaty rights; the nature and main elements of consensus decision-making; the alternative meanings of the terms "co-operative management" and "co-management"; internal conflicts at the heart of both concepts and a way out of the conflict; cooperative management as consensus decision-making within existing legal and land claims authorities; the specifics of the Saoyú-ʔehdacho Agreement and related legal issues; the innovation at the heart of all three agreements; and the essential elements of consensus-based cooperative management agreements and the practice of cooperative management. Using the model set out in these agreements, we can negotiate protected area and cooperative management agreements that reconcile the objectives and terms of modern day land claim agreements and legislation with the older aboriginal tradition of consensus decision making.

T.H.D. Nesbitt (✉)
3377 West 38th Avenue, Vancouver, BC V6N 2X8, Canada

Tuktut Nogait Management Board, P.O. Box 91, Paulatuk, NT X0E 1N0, Canada

Saoyú-ʔehdacho Board, P.O. Box 164, Déline, NT X0E 0G0, Canada
e-mail: tom.nesbitt@telus.net

T.M. Herrmann, T. Martin (eds.), *Indigenous Peoples' Governance of Land and Protected Territories in the Arctic*, DOI 10.1007/978-3-319-25035-9_3

Keywords Protected area agreements • Northern Canada • Reconciliation • Consensus decision making • Cooperative management • Co-management • Aboriginal and treaty rights

1 Purpose and Approach

My purpose in this Chapter is to suggest a different way of thinking about—and practicing—cooperative management. I characterize this style of cooperative management as consensus decision making within existing legislative and land claim authorities. I will examine it through a discussion of three protected area and cooperative management agreements: the Gwaii Haanas Agreement (1993), the Tuktut Nogait Agreement (1996) and the Saoyú-ʔehdacho Agreement (2008).

The focus of the paper is on the cooperative management of protected areas in Canada's north. My objectives include removing unnecessary conflicts in the concept and its practice; maximizing cooperation in cooperative management, given existing legislative and treaty authorities; and contributing to the reconciliation of the interests and worldviews of aboriginal and non-aboriginal Canadians.

Cooperative management can be thought of as an ecosystem of many considerations and relationships. In my experience as a facilitator, negotiator and implementer of protected area agreements and other multi-party land use planning exercises, we will have difficulty understanding and changing the practice of cooperative management if we are unable to see this larger system, its elements and how they function.

In Parts 2, 3, and 4 of this Chapter, I sketch the outlines of this larger system. Subsequently, Parts 5, 6, 7, and 8 provide examples and specifics. The paper is structured as follows. I will:

- briefly describe the purpose of the recognition of aboriginal and treaty rights in Canada's constitution—reconciliation;
- describe the main elements of consensus decision making;
- discuss three models of cooperative management, while recommending a model that imports consensus decision making into cooperative management;
- describe the consensus decision making model as institutionalized primarily in the Saoyú-ʔehdacho Agreement, but also in the Tuktut Nogait and Gwaii Haanas Agreements;
- discuss a case in the Federal Court of Canada that upholds the consensus-based model;
- identify what is innovative about this model; and
- suggest the essential elements of consensus-based cooperative management agreements and the practice of cooperative management.

2 Context: Reconciliation and Land Claim Agreements

Aboriginal and treaty rights, including the rights set out in modern day land claim agreements, are recognized and affirmed in s. 35 of Canada's Constitution Act, 1982. These rights have thus become part of the fabric of Canada. Our constitution is, legally, a living document that must evolve to reflect our changing values and aspirations as a people (Edwards 1929).[1] In my opinion, aboriginal and treaty rights are thus now part of what it means to be a Canadian.

The Supreme Court of Canada has now taken the principles set out in the previous paragraph one step further. In several judgments, it has now informed us that the underlying purpose of modern day land claim agreements, and of the recognition of the rights set out in them, is the reconciliation of the interests and world-views of aboriginal and non-aboriginal Canadians. In Van der Peet (1996), the Court identified reconciliation as the underlying purpose of section 35(1) of the Constitution Act, 1982. At para. 31,[2] it said:

> ... what s. 35(1) does is provide the constitutional framework through which the fact that aboriginals lived on the land in distinctive societies, with their own practices, traditions and cultures, is acknowledged and reconciled with the sovereignty of the Crown. The substantive rights which fall within the provision must be defined in light of this purpose; the aboriginal rights recognized and affirmed by s. 35(1) must be directed towards the reconciliation of the pre-existence of aboriginal societies with the sovereignty of the Crown.

Similarly, in Mikisew Cree (2006), the Supreme Court of Canada said, at para. 1:

> The fundamental objective of the modern law of aboriginal and treaty rights is the reconciliation of aboriginal peoples and non-aboriginal peoples and their respective claims, interests and ambitions.

To reconcile is to make consistent, to respect and heal differences, to seek and find ways of making different concepts or interests compatible, and to ensure that different parties' interests are simultaneously met. The Concise Oxford Dictionary defines "reconcile" as "heal", "settle", ... "harmonize, make compatible, show compatibility of ... (apparently conflicting facts, statements, qualities, actions ... with or and another)". To reconcile is to find ways of accommodating ..., in the sense of finding a place in the house for both, not compromising one for the other. We can reconcile concepts (the tendency of Van der Peet) or interests (the formulation in Mikisew). To reconcile the interests of aboriginal and non-aboriginal Canadians is thus to find means through which their several respective interests are elicited, defined and simultaneously met. And to reconcile the concepts and world-views of aboriginal and non-aboriginal Canadians is to understand that they are complementary, and to find means through which they can co-exist and inform our actions.

This paper looks at cooperative management as a process through which the Crown and aboriginal authorities may define their several respective interests and

---

[1] "The *British North America Act* (the *Constitution Act* of 1867) planted in Canada a living tree capable of growth and expansion within its natural limits." Per Lord Sankey, *Edwards v. Canada* (A-G) (1929), [1930] p. 136.

[2] See also paras. 3, 4 and 43.

make every effort, through consensus, to reconcile these interests (Nesbitt 2006). *Final* reconciliation is unachievable. Accordingly, in Mikisew Cree and Haida Nation (2005), the Supreme Court has spoken of "advanc[ing] the process of reconciliation", of "the goal of reconciliation" and of the "overarching objective of reconciliation".[3] I suggest that what the Supreme Court is getting at in these cases is the never-finished task of reconciliation. It may be helpful to understand the concept of reconciliation as having some parallels with the concept of authenticity in existential philosophy. It is always "an issue".[4] It can never be resolved or achieved in the sense that we can consider it "accomplished", "done", a file we can now put on the shelf and move on. Reconciliation is an ongoing relationship and task between the Crown and Canada's aboriginal peoples, and it must be continually re-done. Our actions can only contribute to an ongoing process of reconciliation.

Finally, reconciliation must, in the Crown-aboriginal context, be closely associated with the concept of mutual respect. Reconciliation and mutual respect are related or "sister" objectives. While we may not always be able to advance both reconciliation and mutual respect, that is surely our ultimate objective, and the process of advancing reconciliation is commonly also one of advancing mutual respect. It is no accident that the Supreme Court has juxtaposed these concepts in Mikisew at para. 49:

> There is in the Minister's argument a strong advocacy of unilateral Crown action (a sort of "this is surrendered land and we can do with it what we like" approach) which not only ignores the mutual promises of the treaty, both written and oral, but also is the antithesis of reconciliation and mutual respect.

Beginning in 1984, various aboriginal authorities and the government of Canada have concluded several modern day land claim agreements in the Northwest Territories, Nunavut and the Yukon. These include the Inuvialuit Final Agreement (1984) and the Sahtu Dene and Metis Comprehensive Land Claim Agreement (1993). These and other land claim agreements profoundly affect the protected areas that fall within their land claim settlement areas. Three principles are worth underlining here.

Land claim agreements are about long-term relationships—and developing and adapting these relationships to new circumstances over time. These agreements create rights, in the beneficiaries to the agreements, to harvest, to land, and (most important here) to participate in resource management decision making.

Below, I discuss the Saoyú-ʔehdacho Agreement, the Tuktut Nogait Agreement and the Gwaii Haanas Agreement. I will readily admit that these agreements are explicitly not themselves land claim agreements. But the former two agreements, and the cooperative management relationships they create, take place and function within the wider sphere of the new relationships created by overarching land claim agreements.[5] The rights created by land claim agreements extend throughout protected areas within claims settlement areas. Thus, in my view, the purpose of recon-

[3] *Mikisew* at paras. 4 and 50; *Haida* at para. 35.

[4] Martin Heidegger, *Being and Time*, translated by John Macquarrie and Edward Robinson (New York and Evanston: Harper & Row, 1962) at 68.

[5] In the case of the *Tuktut Nogait Agreement*, *The Inuvialuit Final Agreement* (1984); and in the case of the *Saoyú-ʔehdacho Agreement*, the *Sahtu Dene and Metis Comprehensive Land Claim*

ciliation must animate not only the applicable land claim agreements, and the relationships they establish, but also the cooperative management relationships established by the former two agreements. Further, since the objective of reconciliation extends not only to treaty but also to aboriginal rights, this objective should properly animate cooperative management relationships throughout Canada, including those established by the Gwaii Haanas Agreement.

The cooperative management of northern protected areas is therefore potentially a mechanism for reconciling the interests of aboriginal and non-aboriginal Canadians over time. It is a means for enriching the cooperative decision making and the day-to-day work of the representatives of aboriginal authorities and the crown. It is also, in my view, potentially a way whereby the crown can more effectively consult and accommodate the interests of aboriginal peoples whose traditional territories include what are now protected areas.

3 The Concept and Practice of Consensus Decision Making

The aboriginal elders with whom I have worked tell me that consensus decision making is an aboriginal concept and practice.[6] They tell me that consensus was traditionally a way of developing and adapting long-term relationships within and between aboriginal collectives. Groups of people would periodically meet, discuss issues and make decisions by consensus.[7]

This paper describes consensus decision making as I understand it to have been traditionally practiced by Inuvialuit and Dene elders, and as subsequently adapted and practiced by the Tuktut Nogait and Saoyú-ʔehdacho Management Boards.

Aboriginal cultures are primarily oral, not written, cultures. Consensus is thus about how we comport ourselves: how we talk to each other, listen to each other, treat each other, and make mutual decisions. It is not just about the words written on a page or a formal (written) agreement. It extends beyond written agreements and is about how we act, how we define ourselves by our actions, and our ongoing relationships with others.

Let me touch here on several aspects of the concept and practice of consensus decision making:

It is about making sound decisions. Many things are going on in consensus decision making. They may not all be explicit, or explicitly stated. But at the end of the

---

Agreement (1993). The *Gwaii Haanas Agreement* (1993) has no associated treaty or land claim agreement, but is subject to aboriginal rights.

[6] I am particularly indebted here to the late Billy Day (Inuvik), who taught me the basic process of consensus decision making. Billy Day, personal communications (1986 & 1987).

[7] Charlie Neyelle, Alfred Taniton and Leon Modeste (all of Déline, NWT), all personal communications (2000–2005). From 2000 to 2005, these three elders worked with me together, as one collective advisor, and they gradually revealed their understanding of consensus and its cultural context. I want to respect this collective cultural approach and thus do not separate these three people or identify a specific date for their personal communications.

day, when we reflect on and analyze what takes place in consensus decision making, we can see that it is fundamentally about sound group decision making. Whether implicitly or explicitly, consensus is a discussion and decision-making process that takes the time needed to elicit and define participants' several individual and common interests; that understands the facts from different perspectives; that identifies and evaluates, where necessary, common options; and that accommodates different parties' interests and perspectives in mutual decisions. Consensus decision making is not, as the word is sometimes used, a vague, majority-based ("most of us agree") way of making decisions. In consensus, participants discuss a family of issues; develop and frame a consensus on them; and agree to/verify that consensus.

Consensus decision making is fundamentally about mutual respect.[8] It's about how we see and treat others and ourselves. It's about taking time, listening and recognizing the different perspectives at the table; giving participants a space to speak and be heard; offering each person an opportunity to speak; and weaving participants' perspectives together into one coherent whole.

The long-term relationship of the parties, and their responsibilities to the future, seem frequently to be present in consensus discussions. The awareness of these things may be implicit and not needing to be stated—or (alternatively) explicit and stated. Participants, particularly elders, will often make comments to the effect that we act now for future generations. Others will comment to the effect that "This Board will be here in 100 years, long after we are gone: what we are doing now will affect future generations." In both cases, I understand participants to be speaking as much about relationships as about responsibilities: our responsibilities require us to attend to our relationships as participants in the consensus decision-making process.

The cadence of consensus-based discussion is important. Participants are allowed to listen and reflect. Spaces are allowed in the discussion. Participants are allowed

[8]What do we mean when we say that we "respect" another? I understand respect according to Kant's famous dictum (1785, trans. 1969, p. 54): "… Act so that you treat humanity, whether in your own person or in that of another, always as an end and never as a means only" — or not only as a means to our own ends, but as ends in themselves. We then *see* others (and ourselves) as ends in themselves. This is admittedly a western philosophical formulation. But in my experience, something like it also appears to be implicit in the approach of the aboriginal elders with whom I have worked. This should not, on reflection, be particularly surprising: while variously formulated, the basic precept is implicit in all of the world's main spiritual traditions. Furthermore, Kant himself insisted that he had nothing new to say about morality, and sought only to make explicit the deepest convictions of the ordinary person of conscience (Wolff, R.P. in Kant, p. vii). If we take this one step further, we can ask where the precept of treating others and ourselves as ends in themselves might lead us? I suggest that we can never *know* in advance where this precept will lead us. These ends cannot be known in advance. They are to be determined. To treat others and ourselves as ends in themselves implies our mutual self-determination. Reconciliation — as accommodation, as respect, as end in itself, as self-determination — is to be determined on an ongoing basis and is a never-finished task. These concepts are part of an interrelated family or constellation of concepts that function together. In the context of consensus decision making, reconciliation and mutual respect are about recognizing the need for others to participate or share in decision making. Without the participation of Canada's aboriginal peoples in decision making, there will never be real reconciliation.

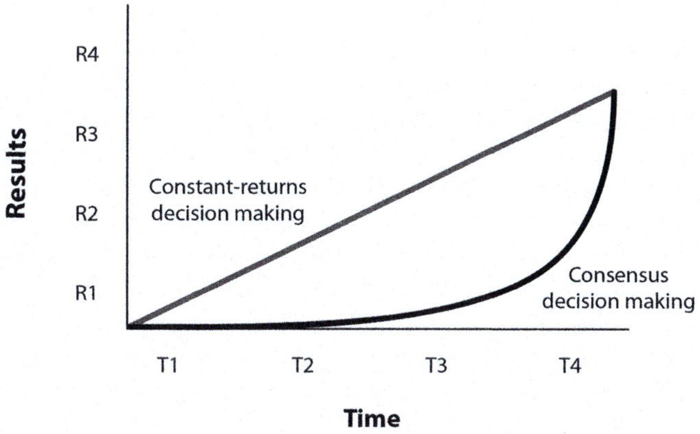

Fig. 1 Decision making, time and results

to try out different perspectives, learn and change their perspectives. No one is held to a position s/he may have advanced earlier in the discussion. Commonalities are identified and different perspectives are integrated or accommodated. At appropriate places in the discussion, the chair or facilitator will formulate, integrate, state and test with participants what s/he takes to be the consensus (including the elements of the consensus) to that point of the discussion.

We often hope or expect to see progress in our discussion as ongoing, linear (Fig. 1). If we were to graph progress against time, we would often like to see progress as a straight-line development. At time 1, we would like to see progress 1, and at time 2, progress 2 … Consensus decision making is commonly not like that. Participants may discuss the issues for some time with little apparent progress. But near the end of the discussions, a consensus will quickly emerge and the different threads of the discussion can be integrated into one coherent whole. Progress is thus typically an exponential curve, not a straight line. Note: Contrary to a popular assumption, I am aware of no objective evidence that would establish or even suggest that the total time required for consensus decision-making processes is greater than for alternative processes. My experience facilitating consensus decision-making processes suggests that consensus decision-making processes are as efficient as any alternative, while the results are typically longer lasting.

Consensus must always be adapted to context, including the cultures at the table and the place in time that participants are at in their discussions. The process should allow participants and their relations with each other to co-exist and to co-evolve. The process, agenda and relationships of month 1 are not the same as the process, agenda and relationships of month 24, as participants evolve a different "culture of the table".[9]

[9]An illustration may help: From 1995 to 2000, I facilitated the work of a "Research Advisory Committee" in Wood Buffalo National Park. It was a "*research* cooperative management committee". The Committee's job was to recommend a research and monitoring program to better understand the incidence and effects of brucellosis and tuberculosis, and other factors such as wolf

Consensus decision making is more about accommodating[10] differences than it is about negotiating compromises among them. While compromises must sometimes be made in discussions and negotiations, it is often assumed that all or most progress in difficult discussions must involve compromise. My experience is quite different. Consensus involves looking for, eliciting and defining each party's underlying interests—and the parties' common interests. Acting from this basis, we can frequently fashion solutions that accommodate these different interests, without needing to make compromises among them. In this, consensus decision making is similar to interest-based negotiations—and the role of the facilitator of consensus discussions is similar to that of the mediator of interest-based negotiations.

In larger groups of people, and in two of the agreements described below,[11] a chair or co-chairs are a negotiated requirement of the consensus decision-making process. The roles of the facilitator/chair in consensus decision making are several-fold:

The facilitator helps establish the foundation—or the ground rules and procedures, whether stated or unstated—of the discussions: mutual respect, the space for all to speak, listening, non-interruption. S/he also helps establish the cadence of the discussions. In this, the facilitator again shares many functions with modern-day mediators. But the facilitator of consensus processes typically has fewer opportunities to intervene than many mediators. The facilitator thus frequently intervenes non-verbally. S/he may prepare an agenda and state rules as needed, but will often model unstated rules by his/her comportment rather than by stating rules.

The facilitator helps ensure that all perspectives and facts are considered and incorporated into the consensus. S/he gives all participants an opportunity to speak,

predation, on bison populations in the Park. The Committee consisted of four aboriginal appointees, three scientists/appointees of government, and one appointee of an environmental non-government organization. The method of investigating and testing hypotheses—the culture—of the three scientists was what Karl Popper (1964) has termed "conjecture and refutation": a vigorous interchange of ideas that tests hypotheses by attempting to refute them—a rapid, spirited but ultimately good-natured back and forth interchange. We cannot finally prove a scientific hypothesis. Rather, the hypothesis that can withstand attempts to refute it is the more likely to be true. In contrast, the method—or culture—of investigating and understanding among the aboriginal representatives on the Committee was one of looking at the applicable traditional knowledge within a consensus decision-making framework. At one of the Committee's early meetings, it became apparent that, unbeknownst to the scientists, their method of investigation and testing ideas excluded the aboriginal representatives. When the Committee became aware of this, it agreed to try out, and soon accepted, the consensus method of proceeding. The scientific and traditional knowledge perspectives could both be accommodated within the consensus decision-making process. Participants' ongoing verbal evaluations suggested that the results of the process were the richer for the interweaving of perspectives, and that the more reflective manner of speaking demanded more of participants and allowed better mutual understanding among them. Relationships among participants evolved. The Committee developed a consensus on and recommended a comprehensive research and monitoring program within a few meetings and the first year of its operation. And it oversaw the research that could be funded within the available budget during the remaining 4 years of its work. Expected conflicts among participants were resolved.

[10] "Accommodating" as "finding accommodation, a place or room for ...".

[11] The *Saoyú-ʔehdacho Agreement* and the *Tuktut Nogait Agreement*.

and at times asks participants each to speak. S/he listens; asks investigative questions; elicits and makes interests explicit as these interests emerge; helps find and identify common interests; helps weave together or integrate the elements of the consensus; and verifies the consensus with participants.

In carrying out all of the foregoing, the facilitator must remain impartial. As in mediated negotiations, the facilitator of consensus decision-making processes tends to oversee the decision-making process while leaving all or most substantive decisions with participants.

In sum, consensus is a process of sound group decision making. It is not a vague-majority based process. It is a process animated by a culture of mutual respect, a more reflective cadence, the space for all to speak, in which perspectives and conclusions can be integrated into a coherent whole, and where participants ultimately verify the resulting whole/consensus.

In this paper I examine cooperative management as consensus decision making within existing legislative and land claim authorities. I want thus to bring consensus decision making as discussed above into the practice of cooperative management.

4 What Does "Cooperative Management" Mean?

There are now many "cooperative management" agreements in northern Canada, for a variety of situations: federally protected areas (national parks, national wildlife areas, migratory bird sanctuaries and some national historic sites), wildlife management, and land use planning. At times, the instituting agreements are incorporated into land claim agreements, and at other times they are not part of but function within the context of land claim agreements.[12] My focus here is on cooperative management agreements for protected areas. But I'd also like to stimulate discussion of the concept and its practice outside of protected areas and outside of land claim settlement areas.

One clarification at the outset: some northern practitioners speak of "co-management". Others insist that "cooperative management" is different. I don't find this distinction helpful. It does not describe a difference in how the terms are actually used in common practice. Nor does it describe an accepted difference in the relationships among the parties established by the allegedly different sorts of agreements. There is, in my opinion, no significant difference in meaning between "cooperative management" and "co-management" as these terms are used in various northern protected area agreements and most other land claims based co-management processes. For the sake of brevity, I use the term "cooperative management" throughout most of the remainder of this paper.

But while there is no significant difference between the meanings of the terms "cooperative management" and "co-management", both terms are sometimes prob-

[12] The *Tuktut Nogait* and *Saoyú-ʔehdacho Agreements* both fall into this latter category. In the case of the Haida and the *Gwaii Haanas Agreement*, no treaty has been settled.

lematic. Both terms can mask differences in parties' understanding of the management relationship between them. The problem lies in different speakers using the same term, whether "cooperative management" or "co-management", but meaning very different management relationships among the parties. This "same term—different meanings" divergence can engender misunderstanding and conflict. My purpose here is to suggest a way out of this conflict and a more enriching management practice.

I suggest that we see the differences in what "cooperative management" (or "co-management") means as a spectrum of possible meanings. This spectrum has two alternatives at either end, and a third, middle alternative between the other two. I will describe all three alternatives, and will evaluate each of them according to two tests:

Test 1: Is the alternative consistent with the legislation by which the protected area is established and managed, and by which the legislature has delegated various responsibilities to the executive?

Test 2: Does the alternative advance cooperation and reconciliation (again, the fundamental objective of the modern day law of aboriginal and treaty rights)?

ALTERNATIVE 1: The Community Committee—Government Consults and Decides: At one end of the spectrum, government: (a) consults community representatives (typically a community protected area committee) and/or one or more management bodies; and (b) makes unilateral management decisions when and as it deems appropriate. This might be termed the old paradigm of community consultation. It harkens back to the Northwest Territories of the 1970s. But it is also, for want of a better alternative, often still the default assumption of some practitioners.

Test 1: Is this alternative consistent with existing legislation? Yes: this alternative is strong on this test. The minister retains the responsibilities given him/her by the legislature.

Test 2: Does this alternative advance cooperation/reconciliation? In my opinion and experience, we must answer this question negatively. Any advances in cooperation and reconciliation are relatively weak, neutral or, more commonly, negative.

Conclusions: Alternative 1 is legally sound but weak, neutral or negative on cooperation and reconciliation. In fact, Alternative 1 often engenders conflict. Northern aboriginal practitioners often express strong frustration at the fact that they have discussed an issue at length with government representatives only to be required, at the end of the day, to await the decision of a distant minister of the crown who was not directly party to the discussions.[13]

[13]An example may help elucidate the point. In the Yukon North Slope Conference, Whitehorse, Oct. 2–4, 2012, several aboriginal representatives on land claim based cooperative management bodies expressed strong frustration that they had worked with government representatives on an important issue for several months, only to be required, at the end of that process, to send their cooperative management committee's recommendations to a minister for his/her ultimate decision

ALTERNATIVE 2: the Regulatory Board: At the other end of the spectrum, the legislature could authorize an independent regulatory board to undertake cooperative management and make relatively final protected area management decisions.

Test 1: Is this alternative consistent with existing legislation? No. This level of delegation by the legislature does not, in my experience, currently exist for northern protected areas. New legislation would be necessary.[14] And such legislative changes are, in my opinion, highly unlikely. Parliament is highly unlikely, for example, to amend the Canada National Parks Act. But this is in effect the alternative I have heard urged by many who are unhappy with the current situation and who want to change it—to accord, for example, with international principles.

Test 2: Does this alternative advance cooperation and reconciliation? The alternative might establish strong cooperative relationships and real reconciliation. On the other hand, however, with the stakes in the final decisions now higher, the norm of Privy Council appointments to regulatory boards, and the norm of decisions by a majority of votes, Alternative 2 could also lead to polarization and adversity within the appointed board(s), and to a continuation of the frustration aboriginal collectives sometimes feel in the relationships they negotiate, in good faith, with representatives of the crown.

Conclusions: Alternative 2 is, in my opinion, highly unlikely in the current legal and institutional culture in Canada. It might establish strong cooperative relationships and contribute to reconciliation. Alternatively, it could lead to polarization, conflict and frustration. In any case, I hope to demonstrate that Alternative 2 is ultimately unnecessary.

It is sometimes believed that there is no middle ground between the two alternatives for cooperative management set out above. Government and non-government participants alike are often, in my experience, internally conflicted, and unintentionally blind to any middle alternative, because they cannot *conceive* how such an alternative could exist. They see a strong division between the decision making role of a minister and the roles of an advisory committee, and they have difficulty bringing these roles together. I have frequently witnessed practitioners wrestling with variations on the following implicit reasoning: "If we are

(Wildlife Management Advisory Council (North Slope) and Yukon Environment 2012). In my opinion, the problem here is often one of government representatives' authority, and the willingness of government departments to send authority to the cooperative management table. The problem could often be remedied by the ministers who have ultimate decision-making authority delegating some of that authority to their representatives at the cooperative management table and by working by consensus.

[14] Protected area legislation can be amended by the terms of modern day land claim agreements and their ratifying legislation. For example, section 12 of the *Inuvialuit Final Agreement*, together with the *Western Arctic (Inuvialuit) Claims Settlement Act*, establish a special conservation regime throughout the Yukon North Slope. They effectively amend existing legislation applicable to the area, including the then *National Parks Act* (now the *Canada National Parks Act*). But once the modern day land claim agreement is signed and concluded, the opportunity to amend existing protected area legislation is largely past.

not dealing with an Alternative 2 board—a legislatively-authorized regulatory board—we must be dealing with Alternative 1 committee. This committee can only be advisory to government. And advisory bodies, which government consults, must be on the outside of, or separated from, management decision making. The minister must decide here, not this committee". I believe that these practitioners would often like this situation to be different. They don't want to work in a conflict situation. But they can often see no alternative to it. If practitioners can see—and understand—no middle ground, they must live by the relationships of Alternative 1, and conflict and frustration often follow.

ALTERNATIVE 3, Consensus Decision making Within Existing Legislative and Land Claim Authorities: There is, in fact, a third, alternative middle ground between the two alternatives above. As stated at the outset of this paper, I define this third cooperative management alternative as consensus decision making within existing legislative and land claim authorities. The agreements that establish Alternative 3 establish management boards that are, from a legal perspective, ultimately advisory (to ministers of the Crown and land claim authorities). They do not—and they cannot—illegally fetter ministers or land claim authorities. In these protected area and cooperative management agreements, the minister has not illegally "passed off" his/her responsibilities (delegated to the minister by the legislature) to a third person or board. The decision-making process of Alternative 3 is consistent with existing legislation, final ministerial decision making within the minister's area of responsibility, and the accountability of ministers to the legislature (and the courts). This decision-making process is also consistent with final aboriginal decision making within areas of aboriginal decision making. But at the same time, the decision-making process yields final, cooperative management board decisions, by consensus among representatives of the authoritative decision-makers on the board, including representatives of the minister (typically the superintendent of a national park) and representatives of the relevant aboriginal authority(ies).

In my experience, most practitioners actually involved in the implementation of Alternative 3 agreements "get" the transition to cooperation and consensus. They say they like working in this context and find it fulfilling. They deal with difficult issues—but the consensus decision-making process, together with the Alternative 3 agreement, gives them a way of resolving issues and advancing reconciliation.

In Alternative 3 protected area and cooperative management agreements, government and aboriginal authorities agree to: manage the protected area cooperatively, according to mutually-agreed management purposes or principles; bring all protected area management issues to the management board forum; send authority to the management board table—both aboriginal authority and ministerial authority; and make all reasonable efforts to make management decisions by consensus, at board meetings.

Test 1: Is this alternative consistent with existing legislation? Yes, consistency is strong. Existing legislative—and land claim authorities[15]—are respected.
Test 2: Does this alternative advance cooperation and reconciliation? Yes, cooperation is strong; reconciliation is advanced.
Conclusions: Both tests above are met. The management board is the place where authoritative parties meet and make every effort to manage by consensus.

The minister/superintendent and authorized aboriginal decision-makers retain their discretion to decide as they see fit. Their discretion is not fettered. Their responsibilities have not been passed off to others. The superintendent can agree or disagree with the consensus. Similarly, if the exercise of land claim rights is at play, the aboriginal representatives on the management board can agree or disagree with the consensus.[16]

We can thus negotiate protected area and cooperative management agreements that integrate the purposes and terms of modern day land claim agreements and legislation with the older aboriginal tradition of consensus decision making. With authority at the table and the wisdom of the consensus decision-making process, we can weave together a "Yes" and contribute to reconciliation.

Thus "advisory boards", ministerial and aboriginal discretion and final, consensus-based decision making need not be mutually exclusive. They can co-exist. There is no need, with mutual understanding and good will, to amend existing legislation or land claim agreements.

5 Three Protected Area Agreements

There are, to my knowledge, at least three protected area/cooperative management agreements in Canada that establish Alternative 3: the Saoyú-ʔehdacho Agreement (2008), the Tuktut Nogait Agreement (1996), and the Gwaii Haanas Agreement (1993).[17]

[15] Or, in the case of the *Gwaii Haanas Agreement* discussed below, where no treaty has been settled, *aboriginal* rights.

[16] Examples: Land claim agreements typically recognize rights to harvest that include the right to establish camps for the purpose of harvesting. The consensus decisions of Alternative 3 management boards are subject to final aboriginal decision making on the exercise of these rights. Similarly, with the inclusion of Sahtu lands within Saoyú-ʔehdacho National Historic Site, discussed below, the consensus decisions of the Saoyú-ʔehdacho Management Board are subject to final Sahtu decision making as regards the exercise of the landowner's rights.

[17] There is a fourth northern protected area cooperative management agreement that readers may find interesting: *Inuit Impact and Benefit Agreement for National Wildlife Areas and Migratory Bird Sanctuaries in the Nunavut Settlement Area* (2006). It is the most comprehensive of the agreements identified in this paper: it covers all of the national wildlife areas and migratory bird sanctuaries in the Nunavut Settlement Area (thirteen protected areas in total); it provides for the establishment of three new national wildlife areas; and it recognizes nine "Area Co-Management Committees". The agreement is, in its structure, an Alternative 3 agreement. In the interests of the

All three of these agreements protect and provide for the cooperative management of places of great value to the aboriginal collectives involved, as well as to Canada as a whole. Aboriginal authorities proposed all three protected areas. The Gwaii Haanas Agreement protects a very significant part of the traditional territory and spiritual and cultural heritage of the Haida. The Tuktut Nogait Agreement protects the calving grounds of the Bluenose barren ground caribou herd, which are of fundamental importance to the Inuvialuit community of Paulatuk as well as to several other northern communities. Finally, the Saoyú-ʔehdacho Agreement protects two of the most sacred places of the Sahtugot'ine (the "people of Sahtu" or Great Bear Lake).

Because it is the clearest and most comprehensive of the three agreements, I will focus here on the Saoyú-ʔehdacho Agreement. But that Agreement is also representative, on the theme of cooperative management, of the relationships established by the Tuktut Nogait and the Gwaii Haanas Agreements.

Saoyú-ʔehdacho is a national historic site on Great Bear Lake, in Canada's Northwest Territories. It consists of two very large peninsulas that reach from the south (Saoyú) and west (ʔehdacho) into Great Bear Lake. It comprises 80 % federally-held lands and 20 % aboriginally-held surface lands. The federal government holds all subsurface lands (or mineral rights) within the site. In total, the site comprises more than 6000 km^2 (slightly more than the size of Prince Edward Island).

The Saoyú-ʔehdacho Agreement, and the management relationships it establishes among its parties, can be seen as the culmination of several agreements. In 1998, with the support of Déline authorities, the minister responsible for Canadian Heritage established Saoyú-ʔehdacho as a national historic site. In 2004, community authorities and representatives of the minister concluded a "commemorative integrity statement" for the site[18]: essentially a statement of what is nationally important about the place (Commemorative Integrity Statement 2004). In 2005, they agreed on a common vision for the site (Nesbitt 2005). In 2007, Saoyú-ʔehdacho was approved by the Northwest Territories Protected Areas Strategy (Sahoyúé-ʔehdacho Working Group, November 27, 2007). And in 2008, the Déline Land Corporation, the Déline Renewable Resources Council and the minister responsible for national parks signed the Saoyú-ʔehdacho Agreement. Pursuant to the Saoyú-ʔehdacho Agreement, the site is legally protected by a subsurface land withdrawal and the transfer of all federal lands to the administration and control of the minister responsible for national parks. The site bears many similarities to a national park.

In the Saoyú-ʔehdacho Agreement, the Déline Land Corporation, the Déline Renewable Resources Council and the minister responsible for national parks (the

disclosure of a potential bias, I should acknowledge that I was one of the chief negotiators of this agreement. In any case, however, since I have no direct experience of the agreement's implementation, I will not discuss it further here.

[18] Interestingly, the site has, from the minister's 1998 declaration, been recognized as nationally significant because of the relationship of Sahtugot'ine and the land, and what we can all learn from this relationship.

"Parties") agree cooperatively to manage Saoyú-ʔehdacho as a whole, disclose all relevant non-confidential information, send decision making authority to the management board table, and make all reasonable efforts to make decisions on the management of Saoyú-ʔehdacho by consensus.

The Saoyú-ʔehdacho Agreement mandates the Parties to establish a six person Saoyú-ʔehdacho Management Board (the "Board"). The Board is basically the place where the Parties meet and manage Saoyú-ʔehdacho by consensus.

The Saoyú-ʔehdacho Agreement establishes a set of common management principles by which the Parties/the Board agree to cooperatively manage Saoyú-ʔehdacho. The Agreement, together with the mutually agreed Commemorative Integrity Statement and the common vision set out in an agreed One Trail Report (Nesbitt 2005), establish the mandate of the Board and the bases for the Parties' management of the site. Parks Canada funds the operation of the site and the work of the Board.

The Board is—as it must be, absent explicit legislation to the contrary—ultimately advisory. Final decision making rests, in areas of federal or territorial jurisdiction, with federal and territorial authorities, principally the minister responsible for national parks. Similarly, final decision making rests, in areas of aboriginal responsibility, with aboriginal authorities: the Déline Land Corporation and the Déline Renewable Resources Council.[19] But the Board is not "only advisory": the Parties have agreed to manage the protected area by consensus. The Board is the place where the Parties bring management issues for final resolution by consensus. It is an Alternative 3 way of cutting between Alternatives 1 & 2 above, and of integrating—and reconciling—the Parties' respective perspectives and interests in managing the protected area over time. There is no term (or end date) on the Saoyú-ʔehdacho Agreement or the work of the Board, although the Parties agree to review the Agreement every 10 years to ensure that it is meeting its agreed principles.

The Saoyú-ʔehdacho Agreement contains provisions for larger issues to include or be referred to the more senior representatives of the Parties. Such larger issues might include the development and approval of a management plan for the site. In these cases, the Board will typically cast the net of consensus decision making more widely, and institute a process that is more inclusive of a wider family of decision-makers.

Finally, if the Board process is unable to yield a consensus, any Party may elect to decide unilaterally, within the area of its jurisdiction. The Saoyú-ʔehdacho Agreement also contains mechanisms for unilateral Party action in the event of emergencies. Neither of the latter options has been exercised since the Board's inception in 2009.

---

[19] Note that, since this and other protected area agreements do not amend land claim agreements, land claims based management authorities (including environmental assessment and wildlife cooperative management bodies) also sometimes have authority in the protected area. But in the three protected area agreements discussed, the protected area is managed, for all practical purposes, by consensus, through the cooperative management board.

Before proceeding, let me deal with an issue that may now have occurred to some readers who are familiar with the Gwaii Haanas Agreement. Some believe the Gwaii Haanas Agreement a unique agreement: essentially an agreement different in nature than the Tuktut Nogait and Saoyú-ʔehdacho Agreements. These people may question my assertion that all three agreements are Alternative 3 cooperative management agreements. I suggest that we understand the issues here as follows:

The Gwaii Haanas Agreement is indeed a unique agreement, in the sense that it arose from unique and very unusual historical and political circumstances. More specifically, this Agreement arose from strong Haida protests to proposals to log Gwaii Haanas, and from the ensuing national and international attention and support for the Haida. The Agreement also required strong federal action, including federal acquisition of the lands now comprising Gwaii Haanas National Park Reserve from the province of British Columbia. The historic novelty of this situation should not be underestimated. But all three of the agreements discussed here are unique, arising from unique historical and political circumstances. I cannot discuss these different circumstances in any detail here. But the unique circumstances of the Tuktut Nogait Agreement would include the conclusion of the Inuvialuit Final Agreement in 1984, after many years of historic negotiations, the community of Paulatuk's call for the establishment of the national park, and 6 years of complex negotiations from 1991 to 1996. Similarly, the Saoyú-ʔehdacho Agreement arose from the Sahtu Dene and Metis Comprehensive Land Claim Agreement (1993), the long years required to negotiate that agreement, community proposals to protect Saoyú and ʔehdacho, and another 12 years of negotiations, from 1996 to 2008, culminating in the Saoyú-ʔehdacho Agreement.

The Gwaii Haanas Agreement is certainly framed, from a legal perspective, in a unique way. In this Agreement, the Council of the Haida Nation and the Government of Canada "maintain viewpoints regarding the Archipelago that converge with respect to objectives concerning the care, protection and enjoyment of the Archipelago, as set out in Section 1.2 below, and diverge with respect to sovereignty, title or ownership" (Gwaii Haanas Agreement 1993, p. 1). In this sense, the Gwaii Haanas Agreement is certainly fundamentally different than the Tuktut Nogait and Saoyú-ʔehdacho Agreements. But that difference is put into perspective when we see that whereas the Haida have agreed to no treaty with the crown, the Inuvialuit and Sahtu Dene and Metis have agreed, respectively, to the Inuvialuit Final Agreement and the Sahtu Dene and Metis Comprehensive Land Claim Agreement. These larger land claim agreements deal with the issues of divergence in the Gwaii Haanas Agreement, and the Tuktut Nogait and Saoyú-ʔehdacho Agreements can thus focus on the parties' common objectives and their future protected area management relationship.

In my view, we can thus separate and put aside the issues of sovereignty, title and ownership in the Gwaii Haanas Agreement. We can see that Agreement for the management relationship actually practiced on a day-to-day basis by the Council of the Haida Nation and Parks Canada. This is a relationship of cooperation and consensus decision making within the framework of existing legislation and aboriginal rights. The Gwaii Haanas, Tuktut Nogait and Saoyú-ʔehdacho Agreements are thus all agreements that bring the age-old aboriginal tradition of consensus decision making into the management of protected areas. They are all Alternative 3 agreements.

Let me now turn to a case in the Federal Court of Canada that has considers and upholds the legal validity of these sorts of agreement.

6 Moresby Explorers Ltd v. Canada (AG) [2001] 4 F.C. 591

Moresby Explorers Ltd is an ecotourism company that operates a float camp in the vicinity of Gwaii Haanas National Park Reserve, on BC's west coast. In 2001, Moresby challenged the superintendent of Gwaii Haanas' decision to refuse to allocate Moresby "user quota", in the national park reserve, for activities associated with Moresby's nearby float camp. Moresby asked the Federal Court of Canada to review and set aside the superintendent's decision. Moresby Explorers illustrates the dilemma set out in Alternatives 1 & 2 above. It is the only case of which I'm aware that directly discusses this dilemma. It is important here because, in this case, the Federal Court upholds Alternative 3 cooperative management.

In this case, Moresby alleged essentially that the minister had, without legal authorization, given over his decision-making authority to the Archipelago Management Board established by the Gwaii Haanas Agreement. Moresby argued that the minister had thus illegally fettered his discretion. In the alternative, Moresby argued that the superintendent had illegally considered matters beyond her authority. There were thus two main issues in the case. While only the first is relevant to the argument here, I will touch briefly on the second to complete the discussion of how the case was ultimately resolved:

Issue 1: Cooperative Management as Consensus Decision making: In making decisions by consensus, through the Archipelago Management Board, did the minister (represented by the superintendent) illegally fetter his/her discretion—or illegally pass off his/her responsibilities under the Canada National Parks Act to the Board? The Court's short answer here is "No". But note:

The case underlines the courts' reluctance to recognize any agreement, between the executive and an aboriginal authority, that would purport to alter a minister's responsibilities, before Parliament, conferred by that legislature, without the clearest of legislative authorization for that alteration. Very clear legislative authority would be required to establish Alternative 2 above (para. 66 of the judgment). The case thus underlines the difficulties that would be associated with establishing Alternative 2. It is not helpful, in my view, to say or imply that all we need to do to improve the management of Canada's northern protected areas is establish one or more Alternative 2 protected areas. The executive, Parliament and the courts would all need to be persuaded: a difficult task.

More importantly here, however, Moresby Explorers stands for the legitimacy of an agreement that establishes an ultimately-advisory board—the Archipelago

Management Board—and a consensus decision-making process, in which the minister's representative (the superintendent) and an aboriginal authority (the Council of the Haida Nation) participate, and through which they make management decisions by consensus. Alternative 3 above is upheld.

Issue 2: Unreasonable Exercise of Ministerial Discretion: Did the minister exercise his discretion in an unreasonable manner? The Court's short answer: "Yes". Canada ultimately lost the case on this second issue. The superintendent did not exercise her discretion in a reasonable manner: she considered activities outside of Gwaii Haanas National Park Reserve and outside of the legal authority conferred by the Canada National Parks Act. She thus acted without legal authority.

7 Innovation in Cooperative Management Agreements

The members of co-operative management (and "co-management") committees sometimes state their frustration that they work for considerable periods of time on management issues, only to have their resolutions accorded the status of recommendations to a distant minister, who may accept or reject the recommendations and who was not in person party to the discussions. The three agreements discussed above establish a different process of decision making and a different relationship between the crown and aboriginal authorities. In my view, they represent true innovation in cooperative management in Canada. Their main innovations consist in the following:

In the Tuktut Nogait Agreement, the crown and the aboriginal parties agree to manage Tuktut Nogait National Park by a set of common management purposes.[20] Similarly, in the Saoyú-ʔehdacho Agreement, the crown and the aboriginal parties

[20] By way of example, section 2 of the *Tuktut Nogait Agreement* reads:

2. PURPOSES

The Parties share the following purposes in creating the Park:

2.1 To protect the Bluenose caribou herd and its calving and post-calving habitat.

2.2 To protect for all time a representative natural area of Canadian significance in the Tundra Hills Natural Region, and to encourage public understanding, appreciation and enjoyment of the area so as to leave it unimpaired for future generations.

2.3 To enhance co-operation between the Inuvialuit and the Government of Canada and the Government of the Northwest Territories in planning, operating and managing the Park.

2.4 To enhance and support local employment and business, and to strengthen the local and regional economies, while making provision for Subsistence Usage within the Park.

2.5 To encourage greater understanding of and respect for the cultural heritage of the Inuvialuit and the natural environment in which it has evolved.

2.6 To provide a setting in which long-term ecological and cultural heritage research may be undertaken.

2.7 To maintain the ecological integrity of the Park, and thereby contribute to the maintenance of the ecological integrity of the ISR [Inuvialuit Settlement Region] as a whole.

The above purposes shall be reflected in the park purposes and objectives statement in the Management Plan.

agree to manage Parks Canada's largest national historic site—Saoyú-ʔehdacho—by a set of common management principles. And in the Gwaii Haanas Agreement, the crown and the Council of the Haida Nation agree that long-term protective measures are essential to safeguard the Archipelago and to apply the highest standards of protection and preservation. Common purposes or objectives are widely accepted as the foundation of any mediated settlement affecting long-term relationships.

In these agreements, the parties agree to make every effort to manage by consensus. Subsection 5.7 of the Tuktut Nogait Agreement reads: "The Board shall make every effort to reach its decisions by consensus. If it is unable to reach a decision by consensus, then it may decide by simple vote." Since its establishment in 1998, the Tuktut Nogait Board has made all decisions by consensus. Section 4.7 of the Saoyú-ʔehdacho Agreement is slightly different. It reads: "The Board shall make its decisions by consensus."—i.e., the Board may only decide by consensus. Since its establishment in 2007, that Board has always been able to reach a consensus on the issues before it. Section 5.1 of the Gwaii Haanas Agreement reads:

> Deliberations of the Archipelago Management Board on any particular proposal or initiative will strive in a constructive and co-operative manner to achieve a consensus decision of the members, which will be deemed recommendations both to the Government of Canada and the Council of the Haida Nation, by way of referral to their designated representatives, agencies or departments, as deemed appropriate by each party.

Consensus works.

In all three agreements, the parties agree to send decision-making authority to the table. In all three agreements, aboriginal authorities are fully represented on the management board and necessary for a quorum and for consensus. Finally, in all three agreements, the minister's representative, the superintendent, participates in management board meetings and the consensus decision-making process.[21]

Let me clarify some of the legal principles underlying the final paragraph above:

The authority exercised by public servants is the authority of the responsible minister. A public servant has, from a legal perspective, no other authority. When a senior public servant, such as the superintendent of a national park, makes a decision, s/he exercises the authority of the minister—to the extent that the minister has conferred that authority on the official—and the minister has made the decision.

The agreements described above require the minister's representative, the superintendent, to participate in the consensus decision-making process and make every effort—or all reasonable efforts—to find consensus. When a consensus is reached, the minister and the aboriginal parties to the agreement have decided. The consen-

[21] In the implementation of the *Gwaii Haanas Agreement*, the Superintendent and another Parks Canada official have traditionally been appointed members of the Archipelago Management Board. In contrast, in the *Tuktut Nogait* and *Saoyú-ʔehdacho Agreements*, the Superintendent is not formally part of the management board, but s/he *is* formally part of the board / party decision-making process. In both alternatives, authority is at the table.

sus is stated and verified by the representatives of all parties. In this case, there is no recommendation to a distant minister and there is no further ministerial decision to be made. The parties have made their decision and they can move on.

In the event that the parties are unable to come to a complete substantive consensus, they might agree that part or all of the matter can be put off (e.g., for further research or reflection). Partial and/or procedural consensus decisions, and recesses, are thus also options.

8 Suggested Elements of Cooperative Management Agreements

Let me conclude by suggesting several elements of future cooperative management agreements and the future practice of cooperative management in northern protected areas. These suggestions are offered in the expectation that they would be adapted as appropriate to different circumstances:

Parties: The signatories or parties to a protected area agreement should include those organizations with a legitimate role in the management of the area. The number of parties should be kept manageable—able to function as a whole and capable of being represented by a management board.

Advisory Board and Subject to Land Claim Agreements: To protect against legal challenges, the protected area agreement should include explicit language confirming that a management board is ultimately advisory. It is also useful, since the question often arises, if the protected area agreement is made explicitly subject to the applicable land claim agreement, such that the rights set out in the land claim agreement are affirmed to run throughout the protected area.[22]

Management by Consensus: Together with the advisory role described above, protected area agreements should also establish that the parties must, through the board, make decisions on the management of the protected area by consensus, or (alternatively) that they must every effort to make such decisions by consensus. Advisory boards and consensus decision making must be conjoined.

Comprehensiveness and Disclosure of Relevant Information: The agreements should establish that the parties must bring all protected area management issues to the board table, and that they must disclose all relevant non-confidential information.

[22] This "subject to [the applicable land claim agreement]" language is ultimately unnecessary in a protected area agreement. Such agreements are unable, absent explicit authorization to the contrary, to amend a modern-day land claim agreement. Land claim agreements are constitutionally recognized and ratified by federal legislation, and a protected area agreement cannot alter these authorities. In my experience, however, it is useful to have such "subject to" language included in the protected area agreements to allay beneficiaries' ongoing and understandable concerns that they may inadvertently have given up or diminished their land claims rights in signing the protected area agreement.

Authority at the Management Table: The agreements should establish that the parties must send authority to the board table. The superintendent responsible for a national park should be part of the management board or, (alternatively) compelled to attend board meetings and participate in the board consensus decision-making process. Cooperative management will only work—and reconciliation will only be advanced—if there is reasonable ministerial and aboriginal authority at the table.

Credibility and Appointment of Board Members: Board members should be sufficiently credible that the board is, as a whole, able to manage the protected area credibly. I suggest that the test is this: does this group of people have the expertise and judgment to make credible decisions—decisions acceptable to all parties—on the management of this area?

Appointments: In the interests of reconciliation, aboriginal authorities should directly appoint their representatives to cooperative management boards, just as a minister of the crown (or the CEO) appoints his/her representatives. Appointments should be staggered, such that several re-appointments are not simultaneously required.

Interest/Engagement/Ability to Work Cooperatively: Board members must want to find a consensus that weaves together their common perspectives and interests. Their interests and personalities must be engaged in the common endeavour. They must "get" the adjustment to cooperative management—the gestalt from Alternative 1 to Alternative 3, and to making decisions by consensus. And the change must be relatively thorough: not made in one part of a board member's mind, only to be contradicted in another. Board members should have demonstrated cooperation skills. They should be selected accordingly.

Time and Effort: The parties to cooperative management agreements, and their appointed management board members, must also be willing to give adequate time and effort to their collective work. Cooperative management only works if the management board members and parties make it work.

Referrals: For matters beyond the authority of party representatives on a management board, the agreements should also make provision for referrals to higher party authorities, or for the inclusion of comments by such higher authorities.

Common Purposes of Principles: The agreements should identify common, agreed purposes or principles for the management of the protected area. Common purposes are the common ground from which protected area agreements spring. They are also the simplest statement of the mandate of the management board—"what we're about"—and the reference point to which the management board and other representatives of the parties will repeatedly return in making difficult decisions.

Board Procedures: The agreements should also make provision for the management board to determine and make publicly available its procedures.

A Long-Term Relationships: Cooperative management and reconciliation are about long-term relationships. The parties should be—and they should understand that they are—in a long-term relationship: the Tuktut Nogait and Saoyú-ʔehdacho Agreements, and the management boards that they establish, exist in perpetuity.

Board members are typically appointed for terms of 2–3 years, renewable. Continuity in board membership is, in my experience, important.

Communications: Communication to/from the parties and appointed board members should be required and funded.

Budget: Management boards must have a reasonable budget to carry out their work. Without an adequate budget—and budgetary authority at the table—the cooperative management relationship has no means of realization.

Different Perspectives: To be sound, consensus decisions must consider all relevant perspectives. Where "outside" expertise (including "staff" expertise) and perspectives are needed, they should be incorporated into the consensus decision-making process.

Co-Evolution: The participants in the consensus decision-making process will need to explore their respective interests and world-views, develop mutual understanding, learn to work together, analyze issues before them, integrate perspectives and develop consensus—and adapt those decisions over time, as required. Participants will co-evolve; relationships will co-evolve; and roles will co-evolve.

Get Out on the Land: To understand and take part fully in the cooperative management of a protected area, management boards need periodically to get out on the land. Managing solely from a desk would, in my opinion, soon exhaust the relevance of a protected area management board.

In the implementation of the Gwaii Haanas, Saoyú-ʔehdacho and Tuktut Nogait Agreements, party representatives have achieved genuine cooperation. They have done this notwithstanding their differences in age, upbringing, culture, power and authority. Pursuant to these agreements, people—who could not be much more different from each other—work together on the basis of common purposes, principles and/or interests. They expand their understanding of the facts and the options available to them. They integrate solutions that accommodate their common interests and their differences. They listen to each other, learn from each other, co-evolve and make protected area management decisions together. They do this through the age-old tradition of consensus. We're all aware of differences, and of the injustices that have been and continue to be visited upon Canada's aboriginal peoples. However in these three agreements, some modest reconciliation of these differences and injustices has been achieved (Figs. 2 and 3).

Fig. 2 A cultural camp and Saoyú-ʔehdacho Board meeting site in ʔehdacho (2010)

Fig. 3 A Tuktut Nogait Management Board meeting campsite at Uyarsivik Lake in Tuktut Nogait National Park: midnight (2011)

References

Canada National Parks Act, S.C. (2000). c. 32 (Can.).

Commemorative Integrity Statement, Sahyoue (Grizzly Bear Mountain) and Edacho (Scented Grass Hills) National Historic Site of Canada. (2004). Retrieved from, http://www.pc.gc.ca/eng/lhn-nhs/nt/Saoyu-ehdacho/site/cis.aspx

Constitution Act, 1982, being Schedule B to the Canada Act 1982 (U.K.), 1982, c. 11, in The Constitution Acts 1867 to 1982, Consolidated as of April 1, 1996. (1996). Canada, Ministry of Supply and Services Canada (Can.).

Edwards v. Canada (A-G). (1929). [1930] A.C. 124 (U.K.C.A.)

Gwaii Haanas Agreement. Between the Government of Canada, represented by the Minister of the Environment and the Council of the Haida Nation, for and on behalf of the Haida Nation and represented by the Vice President of the Council. (1993). Retrieved from, http://www.archive.ilmb.gov.bc.ca/slrp/lrmp/nanaimo/…docs/appendix_2.11.pdf

Haida Nation v. British Columbia (Minister of Forests). (2005) 1 C.N.L.R. 72 (S.C.C.).

Inuit Impact and Benefit Agreement for National Wildlife Areas and Migratory Bird Sanctuaries in the Nunavut Settlement Area. Between the Inuit of the Nunavut Settlement Area, represented by Nunavut Tunngavik Incorporated, the Kitikmeot Inuit Association, the Kivalliq Inuit Association, the Qikiqtani Inuit Association and the Nangmautaq Hunters and Trappers and Her Majesty the Queen in Right of Canada represented by the Minister of the Environment. (2006). Retrieved from, http://www.tunngavik.com/files/2010/02/inuit-impact-and-benefit-agreement-for-national-wildlife-areas-and-migratory-bird-sanctuaries-in-the-nunavut-settlement-area.pdf

Kant, I. (1785, trans. 1969). *Foundations of the metaphysics of morals*. Trans. Lewis White Beck with critical essays, edited by Robert Paul Wolff. Indiana and New York: the Bobbs-Merrill Company Inc.

Martin H. (1962). *Being and time* (J. Macquarrie & E. Robinson, Trans.). New York/Evanston: Harper & Row.

Mikisew Cree First Nation v. Canada (Minister of Canadian Heritage). (2006). 1 C.N.L.R. 78 (Can. S.C.C.). Retrieved from, CanLII database, http://canlii.ca/t/1m1zn

Moresby Explorers Ltd v. Canada (AG). (2001). 4 F.C. 591. (Can. F.C.) Retrieved from, CanLII database, http://canlii.ca/t/ng3

Nesbitt, T. (2005, December 17). *One trail: Facilitator's report on the Sahyoue and Edacho directions-confirming workshop (November 8–10/05)*. Retrieved from, http://www.google.ca/?gws_rd=cr&ei=–fDVUsi1PJD1oASy2oLICw#q=Sahyoue-Ehdacho+one+Trail+REport

Nesbitt, T. (2006, June). *Reconciliation and consensus: Moving the stone*. Vancouver: The Continuing Legal Education Society of British Columbia.

Popper, K. (1964). *Conjectures and refutations: The growth of scientific knowledge*. London: Rutledge.

R. v. Van der Peet. (1996). 4 C.N.L.R. 177. (Can. S.C.C.). Retrieved from, CanLII database, http://canlii.ca/t/1fr8r

Sahoyúé-ʔehdacho Working Group. (2007, November 27). *Sahoyúé-ʔehdacho Working Group Final Report*. Yellowknife: Northwest Territories Protected Areas Strategy Secretariat. Retrieved from, http://www.nwtpas.ca/areas/document-2007-sahoyue-workinggroup-report.pdf

Sahtu Dene and Metis Comprehensive Land Claim Agreement. (1993). *Comprehensive land claim agreement between Her Majesty the Queen in right of Canada and the Dene of Colville Lake, Déline, Fort Good Hope and Fort Norman and the Metis of Fort Good Hope, Fort Norman and Norman Wells in the Sahtu Region of the Mackenzie Valley as represented by the Sahtu Tribal Council* (Vol. 1). Ottawa: Minister of Indian Affairs and Northern Development.

Saoyú-ʔehdacho National Historic Site of Canada Protected Area and Cooperative Management Agreement. (2008, September 7). *Between the Minister of the Environment, the Déline Land*

Corporation and the Déline Renewable Resources Council. Retrieved from, http://www.google.ca/url?sa=t&rct=j&q=&esrc=s&source=web&cd=2&ved=0CDMQFjAB&url=http%3A%2F%2Fwww.pc.gc.ca%2Feng%2Flhn-nhs%2Fnt%2FSaoyu-ehdacho%2Fsite%2F~%2Fmedia%2Flhn-nhs%2Fnt%2FSaoyu-ehdacho%2Fpdfs%2Fsite_pdfs%2FAgreement%2FS-E%2520NHS%2520Protected%2520Area%2520and%2520Cooperative%2520Management%2520Agreement%2520-%2520English%2520text.ashx&ei=qyrWUpqrKJGCogS_z4LABw&usg=AFQjCNHNuoKx7CFG9ss6xwuKYLUYrSBW0Q&bvm=bv.59378465,d.cGU

The Concise Oxford Dictionary of Current English, 7th ed.

The Tuktut Nogait Agreement: An Agreement to Establish a National Park in the Inuvialuit Settlement Region Near Paulatuk, Northwest Territories. (1996, June 28). Between the Government of Canada, represented by the Deputy Prime Minister and Minister of Canadian Heritage, the Government of the Northwest Territories, represented by the Deputy Commissioner of the Northwest Territories, the Inuvialuit Regional Corporation, represented by the Chairperson, the Inuvialuit Game Council, represented by the Chairperson, the Paulatuk Community Corporation, represented by the Chairman, and the Paulatuk Hunters and Trappers Committee. Retrieved from, http://www.pc.gc.ca/pn-np/nt/tuktutnogait/docs/plan1.aspx

Western Arctic (Inuvialuit) Claim Settlement Act, S.C. 1984, c. 24 (Can.).

Western Arctic Claim, The Inuvialuit Final Agreement. (1984). *Between the Committee for Original Peoples' Entitlement, representing the Inuvialuit of the Inuvialuit Settlement Region and the Government of Canada, represented by the Minister of Indian Affairs and Northern Development.* Ottawa: Minister of Indian Affairs and Northern Development.

Wildlife Management Advisory Council (North Slope) and Yukon Environment. (2012). *2012 Yukon north slope conference proceedings: Pathways to Arctic innovation* (pp. 25–55). Retrieved from, http://www.wmacns.ca/resources/yukon_north_slope_conference/

Part II
Transboundary and Regional Protected Area Governance

Two Parks, One Vision – Collaborative Management Approaches to Transboundary Protected Areas in Northern Canada: Tongait KakKasuangita SilakKijapvinga/Torngat Mountains National Park, Nunatsiavut and le Parc national Kuururjuaq Nunavik

Harvey Lemelin, Margaret Johnston, Dave Lough, Judith Rowell, Wayne Broomfield, Gary Baikie, and Kristie Sheppard

Abstract The recent designations of Tongait KakKasuangita SilakKijapvinga/Torngat Mountains National Park (TMNP) and Kuururjuaq Parc National (KPN) have established Canada's newest polar transboundary protected area (TPA) along the Labrador Peninsula of Northern Quebec and Newfoundland and Labrador. The designations have helped to address concerns raised regarding the isolated nature and fragmented biological distribution, and the susceptibility to anthropogenic disturbances of the area. The adaptive management approach used in this TPA fosters regionally-based approaches to protected area management, and promotes regional collaborative developments. These regional initiatives are facilitated through an all Inuit Co-operative Management Board for the TMNP and a harmonization committee overseeing the management of the KPN. Although the mandate of each park committee is to provide advice and guidance for the management of their respective parks, each has also become an important forum for facilitating more regionally-based management approaches through protected areas. This chapter

H. Lemelin (✉) • M. Johnston
School of Outdoor Recreation Parks & Tourism, Lakehead University, Thunder Bay, ON P7B 5E1, Canada
e-mail: harvey.lemelin@lakeheadu.ca; mejohnst@lakeheadu.ca

D. Lough
Culture, Recreation and Tourism, Torngasok Cultural Centre, Nain, NL, Canada
e-mail: dave_lough@nunatsiavut.com

J. Rowell • W. Broomfield • G. Baikie
Torngat Mountains National Park, Parks Canada Agency, Nain, NL, Canada
e-mail: judy.rowell@pc.gc.ca; torngats.info@pc.gc.ca; gary.baikie@pc.gc.ca

K. Sheppard
SilakKijattisinimmut Aulatsijik; Nunatsiavut kavamanga, Nunatsiavut Government, Nain, NL, Canada
e-mail: kls.sheppard@gmail.com

T.M. Herrmann, T. Martin (eds.), *Indigenous Peoples' Governance of Land and Protected Territories in the Arctic*, DOI 10.1007/978-3-319-25035-9_4

examines how collaborative management strategies have been implemented at this regional level through the help of the Indigenous Stewardship Model applies to the TK-TBA. Special emphasis is placed upon discussing issues related to effective Indigenous involvement in governance and ensuring local economic benefits to Indigenous groups.

Keywords Indigenous Stewardship Model • Inuit • Management • National parks • Tourism • Transboundary protected areas

1 Introduction

The Inuit of Labrador have a 600-year history of interaction with European explorers, whalers, fishers, settlers, missionaries, naturalists, and adventurers that is characterized by an adaptation to changing conditions and adoption of new technologies (Brice-Bennett 1977). Recent interactions have taken place with members of the American military, the resource development workforce and others involved in a range of tourism activities related to the non-aboriginal presence in Labrador (Byrne 2008; Natcher et al. 2012; Rodon and Grey 2009). Much of the current tourism industry in Nunatsiavut (Northern Labrador) and Nunavik (Northern Quebec) focuses on cruise tourism, wildlife tourism and adventure tourism (Hull 1999; Lemelin and Maher 2009; Maher and Lemelin 2011). The following discussion examines the potential role of regional tourism approaches and co-management in Northern Labrador and Quebec.

In an attempt to expand and diversify beyond hunting and fishing expeditions in Northern Quebec, a group of tourism proponents established the Nunavik Tourism Association (NTA) in 1998, one of the first Inuit tourism associations in Canada (Lemelin et al. 2012c). In addition to the establishment of the NTA, the regional Inuit organizations such as the Makivik Corporation and its subsidiaries, the Kativik Regional Government and the Nunavik Parks agency were created to address environmental protection, resource management and economic development needs (ARK 2005; Jacobs et al. 2009; Lequin and Cloquet 2006; Ministère du Développement durable, de l'Environnement et des Parcs (MDDEP) 2007). The co-management of these protected areas (Gendreau et al. 2012; Hébert 2012; Martin 2005; Milo and Lariviere 2012) along with investments in regional infrastructures have provided the opportunity for Nunavik to diversify tourism offerings and they now include an annual musk ox hunt, hiking, mountain climbing, and rafting opportunities (ARK 2005; Girouard 1998; Lemelin et al. 2012c; Martin 2012; Thomas 2012).

The Nunatsiavut Government and the Tongait KakKasuangita SilakKijapvinga/Torngat Mountains National Park (TMNP) were established in 2005 (Parcs Canada 2010), three years later, the Kuururjuaq Parc National (KPN) was established (ARK 2005; MDDEP 2007). The Department of Culture, Recreation and Tourism and

Tourism Nunatsiavut were created in-order to develop, promote and market tourism in Northern Labrador.

As stated earlier, the establishment of these two parks created a large transboundary protected area (TPA) along the Labrador Peninsula of Northern Quebec and Newfoundland and Labrador (Lemelin 2012). These regional initiatives are facilitated through the Co-operative Management Board for the TMNP, made up entirely of Inuit board members, and a harmonization committee overseeing the management of the KPN (Lemelin 2012; Lemelin et al. 2012b). Each park committee is mandated to provide advice and guidance for the management of each respective park, but each also has become an important forum that supports regional management approaches. Researchers suggest that a co-operative management arrangement, where decision-making power is shared between two or more bodies, provides a new form of governance that increases effectiveness, efficiency, and equity (Armitage 2007; Bennett et al. 2012; Berkes 2009; Fugmann 2012; Plummer and Armitage 2007). Since the completion of the land claims processes in 1975 in Nunavik and 2005 in Nunatsiavut, the residents of these regions have shaped their co-operative management of their lands and resources (Gendreau et al. 2012; Milo and Lariviere 2012). The management approach in TMNP and KPN is based on a symbiosis between Inuit knowledge and science, and tourism strategies promoting capacity-building, equity generation, and empowerment (Lemelin and Baikie 2012; Lemelin et al. 2012c; Maher and Lemelin 2011). Such partnerships also promote empowerment and ownership, boost conservation initiatives, increase the flow of resources into protected area management, stimulate innovation and foster collaborative decision making and conflict resolution (Moore and Weiler 2009).

This chapter stems from 7 years of collaborative research conducted with partners in Nunavik and Nunatsiavut. The results of this collaborative effort include community reports, presentations, various publications, and the hosting of the 3rd International Polar Tourism Research Network (IPTRN) Conference, "From talk to action: How tourism is changing the Polar Regions" held in the winter of 2012 in Nain, Nunatsiavut (see the conference proceedings from the 3rd IPTRN meeting in Lemelin et al. 2013). This chapter co-written by community members from Nain (e.g., Parks Canada and Nunatsiavut Government personnel) and researchers, illustrates the ways in which the regional collaborative approach in this transboundary protected area has resulted in regional governance strategies and collaborative approaches to tourism. It also discusses how this particular TPA provides insights into other (existing or proposed) polar TPAs managed partially or wholly with Indigenous partners. The chapter then examines each park in detail and provides an overview of the challenges and opportunities. The next section of the chapter provides a discussion of the ISM and an overview of each territory, our discussion then moves on to theemergence of tourism, and the establishment of protected areas in Nunatsiavut and Nunavik.

2 The Territories of Nunavik and Nunatsiavut

Nunavik comprises 33 % of the landmass of Quebec, but has a population of just over 12,000 people (2011) in 14 villages spread along the coastlines of Hudson and Ungava Bays (Lemelin et al. 2012c). Air and marine transportation systems provide essential goods and services to the residents and are central in keeping the communities connected (Berclaz et al. 2006; Martin 2005). Nunavik was established in 1975 through the Convention of the James Bay and Northern Quebec Agreement between the Aboriginal peoples of Northern Quebec and the provincial government of Quebec (Martin 2005; Rodon and Grey 2009). As stated earlier, this agreement provided the foundation for the establishment of regional organizations such as the Makivik Corporation and its subsidiaries (e.g Air Inuit, First Air), and the Nunavik Regional Board of Health and Social Services. Nunavik has a mixed economy comprised of the public sector, the private sector, the social economy, tourism, and harvesting of renewable resources (Martin 2003, 2005, 2012). Resource development includes mining and hydro-electric sites and is supplemented by tourism, transportation, the service industry, and cooperatives (Jacobs et al. 2009; Winter and Durhaime 2002).

Nunatsiavut, meaning our beautiful land, was established as a political entity in January 2005 (Natcher et al. 2012). The Labrador Inuit Association (LIA), representing approximately 5,300 beneficiaries, the Government of Canada, and the Government of Newfoundland and Labrador signed the Labrador Inuit Land Claims Agreement (Natcher et al. 2012). Before going any further, it is important to note that there are important differences in governance between the territories established by the federal government (i.e., Nunavut, NWT, Yukon) and those established by the provinces of Quebec (e.g., Nunavik) and Newfoundland-Labrador (e.g., Nunatsiavut) (Martin 2003, 2005; Natcher et al. 2012; Rodon and Grey 2009). While all both federal and provincial territories set out details of land ownership, resource sharing, and self-government (Vander Klippe 2008), the relationship between Nunavut, NWT and the Yukon are with the federal government, while the relationship between Nunatsiavut and Nunavik are with the provinces of Newfoundland and Labrador and Quebec respectively, and with the federal government (Lemelin and Johnston 2008). The added layer of government within the provincial territories often exacerbates governance and management issues.

Incorporating the traditional territory of the Inuit of Labrador, the territory of Nunatsiavut is composed of five communities: Nain, Hopedale, Makkovik, Postville, and Rigolet. There are 6,500 Nunatsiavut beneficiaries, many of whom live outside the region. Economic development in Nunatsiavut is largely resource based and is focused around a nickel mine at Voisey's Bay and hydro-electric projects at Churchill Falls (Lemelin et al. 2012b). The Nunatsiavut Group of Companies is the business branch of the Nunatsiavut Government. Its mission is to create wealth in trust for Nunatsiavut Beneficiaries through owning and operating profitable, sustainable businesses, including marine and air transportation, commercial real estate, construction, remote camp operations and logistics. Other economic opportunities include the public and private sectors, harvesting of renewable resources, fishing, cooperative associations, tourism, transportation, and the service industry (Lemelin et al. 2012b).

2.1 Torngat Mountains National Park and KNP

Established as part of the Nunatsiavut land claim in 2005, TMNP extends from Saglek Fjord in the south, to the northern tip of Nunatsiavut and from the watershed boundary in the west to the waters of the Labrador Sea in the east – roughly 9,700 km^2 (Parcs Canada 2010). The management of TMNP is the responsibility of a seven-member co-operative management board made up two members from Nunatsiavut, two from the Makivik Corporation (representing the Nunavik Inuit), two from Parks Canada and one elected by the other three parties to act as Director of the Board. The TMNP co-operative board is the first all Inuit co-operative management board in the history of Parks Canada (Lemelin and Baikie 2012).

Established in May 2009, the KPN is the second park established in Nunavik (the first was Pingualuit Park in 2004) and the 24th park in the network of national parks[1] in Quebec (ARK 2005). At 4273 km^2 KNP protects the watershed of the Koroc River, from its source in the Torngat Mountains to its estuary at Ungava Bay and the spiritual grounds of the Torngat Spirits for which the TMNP is named (MDDEP 2007). The parks also recognize and protect key sites of Inuit culture, including tent circles, sod houses, food caches, burial sites, and *Aullâsimauet* (Inuit settlement camps) (Lemelin and Baikie 2012). Additional cultural sites in the TBPA includes the Moravian missions at Okak and Rama, the Hudson Bay posts at Saglek Bay and Nachvak Fjord, a German meteorological station on the Hutton Peninsula from World War II, and two radar sites dating from the cold war (Lemelin 2012; Lemelin et al. 2012b).

Though the committees are mandated to provide advice and guidance for their respective parks, they have begun to function as well as a means of facilitating a greater focus on regional management approaches (Lemelin 2012). Indeed, by linking protected areas in Quebec and Newfoundland/Labrador within the traditional Inuit territories of Nunatsiavut and Nunavik, this initiative has helped to address concerns raised about the need to protect the environment and the habitat of fragmented populations, and problems arising from anthropogenic disturbances in the area (Lemelin and Baikie 2012; Lemelin et al. 2012b, c). The adaptive management of this TPA also fosters regional approaches to protected area management and promotes regional collaborative developments (Fugmann 2012; Lemelin 2012). The following section examines the impact of the establishment of these two protected areas on two gateways communities, and how these parks have facilitated the establishment of a partnership management approach and the implementation of regional tourism strategies.

[1] The adoption of the Parks Act of Québec (chapter P-9) in 1977 and the significant changes which were made in 2001, including those that eliminate the classifications of "recreation" have allowed include provincial parks in Quebec within the larger international family of national parks under the criteria established the International Union for Conservation of Nature (Lemelin 2012).

3 Regional Tourism Developments

The gateway communities for the park are the two Inuit settlements of Nain in Nunatsiavut, and Kangiqsualujjuaq in Nunavik. Nain is the central administrative centre for TMNP while Kangiqsualujjuaq is the headquarters for KPN. Nain, accessible only by boat or plane, is located on the shores of Unity Bay and is the most northerly community in Nunatsiavut. Nain is also the largest community in the region and has a population of just over 1100, most of whom are Inuit and Kablunagajuit (those of mixed European settler and Inuit descent). The main office for the TMNP and the arts and crafts shop are located within the same building. The headquarters for Parks Canada will be relocated once the Torngâsok Cultural Centre. In addition to housing Parks Canada, the Torngâsok Cultural Centre once completed, is also intended to act as a cultural and interpretive centre for local people and visitors to Nain.

Kangiqsualujjuaq is located near the coast of Ungava Bay at the mouth of the George River and has a population of nearly 900. Most of the local income in the hamlet of is derived from traditional subsistence activities, transfer payments and government employment. The KPN pavilion is staffed by employees of Nunavik Parks and one-part time employee working for the TMNP. Since both parks are located some distance from the communities, they must be accessed with the help of snowmobiles, boats, planes or helicopters. The two gateway communities of Nain and Kangiqsualujjuaq facilitate this east-to-west or west-to-east visitor access.

Just outside of the TMNP's southern boundary on the shores of Saglek Bay is the base camp that functions as a staging point for researchers, contractors, park staff, and other visitors. It is also a registration, education and interpretation venue for tourists on cruises, private yachts and outfitted trips (Maher and Lemelin 2011). The Torngat Mountains base camp and research station was established in part to provide logistical support for research activities and visitors in this remote area. Established in 2006, the first base camp was located within the park on Shuldham Island and comprised a small collection of tents with no electricity and no direct fresh water supply. Shuldham Island was deemed to be inefficient and the base camp was relocated to its present location in St. John's Harbour in 2007. This particular site was selected in part because it had already been used extensively in the 1960s during the construction and maintenance of the Saglek Radar Base and the Saglek air strip. In the 1980s, the site was used by a big game outfitter who built a large cabin that was subsequently destroyed by an avalanche. In 2009 stimulus funding from the Federal Government provided an opportunity to establish permanent facilities at St. John's Harbour. That same year, the operational management of base camp was passed from Parks Canada to the Nunatsiavut Group of Companies. The base camp has been operated by the Nunatsiavut Group of Companies since then.

Each year the camp employs 18–20 full-time staff members (site managers, cooks, maintenance, bear guides, outfitters) who are beneficiaries of the Nunatsiavut Land Claim Agreement or residents of Nunavik. The camp operates primarily on green technology (solar power), but it is also equipped with back up diesel generators. Some of the facilities found at the base camp include a full commercial kitchen, a cafeteria seating approximately 30 people, office and laboratory space, meeting rooms, and washroom, shower and laundry facilities. The Torngat Arts andCrafts organization, based in Nain, operates a craft shop out at the base camp each season with a variety of items for sale and also provides opportunities for visitors to learn more about carving. Tourism and experiential learning opportunities are provided through local guides, which have become a highlight of the Torngat experience (Lemelin et al. 2012c). Visitors can opt to stay in various types of accommodations ranging from insulated, electric-heated designed shelters to backpacking four-season tents. All of these accommodations are located within an electrified fence compound.

Despite the park's distance from major population centres 461 or more tourists have visited the park in each season (from 2009 to 2012) (Brackley et al. 2011). Though these numbers appear low, they are considerably higher than a number of other remote national parks located in Northern Canada (Brackley et al. 2011; Lemelin et al. 2013). Although backcountry use in TMNP is low at present, the establishment of (KPN) to the west will increase access opportunities for canoeists, kayakers and hikers along the Koroc River corridor and in the area of Mount D'Iberville (Lemelin 2012). Visitor numbers for the KPN could not be obtained although it is estimated that a number of fishers, hikers, and white-water adventurists make their way into the park each year (Lemelin 2012). Along with employment opportunities, the two parks also provide opportunities for local Inuit to visit these areas. According to Bennett and colleagues (2012) and Lemelin and colleagues (2012c), these opportunities are essential if local people are going to understand the value of protecting this area and to reconnect to the land and its history.

4 Challenges and Opportunities

Several authors have described the challenges associated with northern tourism initiatives: some of these factors include: remoteness from travelling markets; lack of infrastructure, and high costs (e.g., Johnston 1995; Hall and Boyd 2005; Müller and Jansson 2007). Human-wildlife interactions in these remote areas can also present unique management challenges (Lemelin and Maher 2009). Access to human and economic capital and information is also a major hindrance to tourism and other types of economic development fortourism in peripheral regions (Boyd and Butler 1999). Despite these challenges, the regional partners continue to pursue tourism as a means of development to better their social and economic futures or to diversify regional economies (Fuller et al. 2007). Indeed, the establishment of national parks like KPN in Nunavik and their associated infrastructures (e.g. visitor centres) provide new opportunities to connect or re-connect residents with these areas, address current problems (i.e., spatial disconnection, service capacity, infrastructures), capture user fees from canoeist and other non-consumptive users, all the while implementing tourism diversification strategies (Lemelin 2012). Fostering and promoting new networks and tourism opportunities included diversifying current tourism opportunities through partnerships with private operators (ABV Tours, Arctic

Kingdom, Adventure Canada, Rapid Lake Lodge), local and regional associations (Nunavik Parks, Nunavik Tourism Association, Nunatsiavut Tourism), post-secondary institutes (colleges and universities) and national organisations (Parks Canada), and agreements like the Pan Northern Cooperative Agreement (Lemelin et al. 2012a). Promotion is also an important component of tourism, and the TMNP has been featured on the cover of Newfoundland and Labrador's 2011 tourism guide and was selected as a 'Signature Experience' by the Canadian Tourism Commission (CTC). The CTC created the Signature Experience Collection to identify tourism businesses that can showcase Canada to the world. Next we examine how the Indigenous Stewardship Model applies to the TK-TBA.

4.1 The Indigenous Stewardship Model (ISM)

The ISM integrates cultural and spiritual perspectives with the mixed economy (subsistence, wage and social economies), in order to provide a comprehensive management approach that promotes sovereignty, self-governance and self-sufficiency in Indigenous territories (Kakekaspan et al. 2013; Ross et al. 2011). As table one illustrates, the ISM is founded upon an adaptive and cooperative framework and contains four components: active Indigenous stewardship on traditional territories, community outreach to support Indigenous stewardship, conflict resolution, and co-management (Ross et al. 2011). The ISM is not a rigid prescriptive approach but rather a suggestion "of things to be aware of – recommendations, alternatives, potential solutions, cautionary tales, and encouragement – and . . . warn[s] of pitfalls" (Ross et al. 2011:241). Despite being a relatively new co-management model applied mostly to natural resource management in the USA, and more recently, to polar bear management in Northern Ontario, Canada (Kakekaspan et al. 2013), the ISM was deemed particularly relevant for this case study since it provides a template to develop, implement and monitor co-management strategies in protected area management.

As Table 1 illustrates, the establishment of TMNP and KPN provided a number of benefits including the creation of management boards composed of local people, the establishment of parks research and monitoring strategies combining both Inuit knowledge and scientific approaches, regional tourism strategies guiding present and future tourism development projects in the parks, healing the hurt by bringing local members of both communities back into the TBPA, and the generation of employment opportunities in both the park and in the gateway communities. This case study provides a good baseline description of how the Indigenous Stewardship Model can work in the context of modern treaties and parks systems that emphasize collaborative management approaches in a transboundary context. That said, since both parks and tourism management strategies are early in the development stages and further evaluation documenting the effectiveness of these management strategies beyond this baseline analysis will be required.

Table 1 The Indigenous Stewardship Model and the TK-TBA

Component	Description of Indigenous Stewardship in TK-TBA
A – ***Active Indigenous Stewardship on Tribally Controlled Lands***	
Indigenous ecology: Land and habitat maintenance, restoration, preservation	Recognition of Inuit knowledge and conservation approaches in decision-making.
Subsistence lifestyles: access to and equitable distribution of resources	Recognition of Aboriginal rights through treaties and management agreements.
Promoting economic self-sufficiency (sustainable harvesting and ecologically sustainable micro-enterprise development)	Production and sale of crafts and carvings at the TMNP and in the community of Nain.
Connections to the land: community monitoring and reporting	Reconnecting younger generations with elders out in the parks through employment opportunities and experiential education.
B – ***Community Outreach to support Indigenous Stewardship***	
Indigenous ethnobotany: indentifying, restoring, disseminating Indigenous knowledge	Publishing, presenting and otherwise disseminating traditional knowledge. Incorporating some of this knowledge in educational and interpretation strategies for the parks.
Community input: synergies between Indigenous knowledge and management systems	The collaborative management approaches foster the incorporation of Inuit knowledge and conservation approaches in decision-making processes associated to both parks.
Indigenous pedagogy: Intergenerational transfer of Indigenous knowledge	Reconnecting the younger generations with elders out in the parks through employment opportunities and experiential education.
C – ***Co-management Advocating for Indigenous Stewardship on Land Where Authority is Shared or Absent***	
Validating Indigenous knowledge systems: policy advocacy and reform	Information dissemination through workshops and conferences held in the communities.
Strategies for genuine collaboration: avoiding appropriation and dominance	Co-management committee is to develop best management practices.
D – ***Consensus Building and Conflict Management***	
Indigenous processes for decision making: Resolving internal differences	Using various strategies (i.e., community meetings, workshops, and the web) to inform local communities of upcoming event.
Indigenous collaboration in decision making: Resolving external differences	Developing codes of conducts, appropriate interpretation strategies and conflict resolution strategies

Based on Ross et al. (2011)

5 Conclusion

In this chapter we sought to demonstrate how transboundary protected approaches are closely linked to the Indigenous landscape perspectives. Through a collaborative approach our goal was to demonstrate how establishing transboundary protected areas in Canada's provincial north can protect large ecological zones and key cultural areas, assuring that livelihoods are safe guarded and fostered, implementing research strategies emphasizing both scientific and traditional knowledge systems, and generating regional economic opportunities through tourism. Although transboundary approaches have been used in other situations to promote regional tourism (Hitchner et al. 2010), we recognize that this approach to protecting landscapes and/or seascapes is simply one of many now available to Indigenous peoples in the Canadian North (Lemelin et al. 2012a), and we encourage Indigenous groups to establish protected areas that ensure the maintenance of their livelihoods and that will be most suitable to their needs and values. Developed by and for Indigenous peoples, we believe that the ISM is a valuable instrument providing both baseline figures and future aspirations for the development of protected areas.

Given the early stages of management in the two parks and the existence of two park plans and two managing groups, it will be instructive to continue watching how regional cooperation plays out in the years ahead. The regional gains have been important and recognition of that may provide the impetus for ongoing cooperation despite the jurisdictional challenges in transboundary protected areas.

Acknowledgements Many of the studies cited in this chapter were supported by the government of Nunatsiavut, Parks Canada, Makivik and the Social Sciences Humanities and Research Council of Canada.

References

ARK – Administration régionale Kativik. (2005). *Projet de parc de la Kuururjuaq (Monts Torngat-et-Rivière-Koroc). État des connaissances, Kuujjuaq*, Québec : Administration régionale Kativik, Service des ressources renouvelables, de l'environnement et de l'aménagement du territoire, Section des parcs. 217 p.

Armitage, D. (2007). *Adaptive co-management: Collaboration, learning, and multi-level governance*. Vancouver: UBC Press.

Bennett, N., Lemelin, R. H., Koster, R., & Budke, I. (2012). A capitals framework for appraising and building capacity for tourism development in Aboriginal protected area gateway communities. *Tourism Management, 33*, 4752–4766.

Berclaz, A., David, J., Gosselin, C., Labbe, J. Y., Lacoste, P., & Leclair, A. (2006). Government geoscience stimulates mineral exploration in The Superior Province, Northern Quebec. *Geoscience Canada, 33*(2), 60–75.

Berkes, F. (2009). Evolution of co-management: Role of knowledge generation, bridging organizations and social learning. *Journal of Environmental Management, 90*, 1692–1702.

Boyd, S., & Butler, R. W. (1999). Definitely not monkeys or parrots, probably deer and possibly moose: Opportunities and realities of ecotourism in Northern Ontario. *Current Issues in Tourism, 2*(2/3), 123–137.

Brackley, C., Fick, S., & Greig, K. (2011). À la carte: Alternative dimensions: Resizing Canada's national parks by numbers of visitors. *Canadian Geographic Magazine*, 34–35.

Brice-Bennett, C. (Ed.). (1977). *Our footprints are everywhere. Inuit land use and occupancy in labrador*. Nain: Labrador Inuit Association/Labrador inuit kattekategeninga.

Byrne, A. (2008). *Selling simplicity: Lee Wulff, Stanley Truman Brooks and the Newfoundland Tourist Development Board, 1925–1946, mémoire de maîtrise*. St. John's: Memorial University of Newfoundland.

Fugmann, G. (2012). Developing a remote region: Tourism as a tool for creating economic diversity in Nunatsiavut. *Études Inuits/Inuit Studies, 36*(2), 13–34.
Fuller, D., Caldicott, J., Cairncross, G., & Wilde, S. (2007). Poverty, indigenous culture and tourism in remote Australia. *Development, 50*(2), 141–148.
Gendreau, Y., Gagon, C. A., Berteaux, D., & Pelletier, F. (2012). Cogestion adaptative des parcs du Nunavik dans un context de changement climatiques. *Téoros, 31*(1), 82–91.
Girouard, C. (1998). Le tourisme au Nunavik. *Téoros, 17*(2), 26–27.
Hall, M., & Boyd, S. (Eds.). (2005). *Nature-based tourism in peripheral areas: Development or disaster?* Toronto: Channel View Publications.
Hébert, A. (2012). Les parcs nationaux du Nunavik et du territoire d'Eeyou Istchee-Baie James. *Téoros, 31*(1), 9–18.
Hitchner, S. L., Lapu Apu, F., Tarawe, L., Galih, S., Aran, S. N., & Yesaya, E. (2010). Community-based transboundary ecotourism in the heart of Borneo: A case study of the Kelabit Highlands of Malaysia and the Kerayan Highlands of Indonesia. *Journal of Ecotourism, 8*(2), 193–213.
Hull, J. (1999). Charting adventure tourism development in the North: Labrador's wilderness experience. *Téoros, 18*(2), 15–23.
Jacobs, P., Berrouard, D., & Mireille, P. (2009). *Nunavik: A homeland in transition*. Kuujjuaq, Kativik Environmental Quality Commission. Retrieved from http://www.keqc-cqek.ca/KEQC-AR09-e-F-lo.pdf
Johnston, M. E. (1995). Patterns and issues in Arctic and Subarctic tourism. In C. M. Hall & M. E. Johnston (Eds.), *Polar tourism: Tourism in the Arctic and Antarctic regions* (pp. 27–42). Chichester: Wiley.
Kakekaspan, M., Walmark, B., Lemelin, R. H., Dowsley, M., & Mowbray, D. (2013). Developing a Polar Bear co-management strategy in Ontario through the Indigenous Stewardship Model. *Polar Record, 49*(3), 230–236.
Lemelin, R. H. (2012). L'aire protégée transfrontalière Torngat-Koroc: Gouvernance et retombé économique potentielles dans les territoires du Nunatsiavut (Labrador) et Nunavik (Québec). *Téoros, 31*(1), 29–38.
Lemelin, R. H., & Baikie, G. (2012). Our home and native land: Recognizing the socio-cultural dimensions of last chance tourism. In R. H. Lemelin, J. Dawson, & E. Stewart (Eds.), *Last chance tourism. Adapting tourism opportunities in a changing world* (pp. 168–181). New York: Routledge.
Lemelin, R. H., & Johnston, M. (2008). Northern protected areas and parks. In P. Dearden & R. Rollins (Eds.), *Parks and protected areas in Canada: Planning and management* (3rd ed., pp. 294–313). New York: Oxford University Press.
Lemelin, R. H., & Maher, P. (2009). Nanuk of the Torngats: Human-Polar Bear Interactions in the Torngat Mountains National Park, Labrador, Canada. *Human Dimensions of Wildlife, 14*(2), 152–155.
Lemelin, R. H., Smith, P., & Golden, D. (2012a). Transforming protected areas to support indigenous livelihoods: Examining conservation engagement and responsibilities in two provincial Boreal Norths. In M. Beaulieu & R. Harpelle (Eds.), *Pulp friction: Northern communities in a changing world* (pp. 121–138). Thunder Bay: Centre for Northern Studies.
Lemelin, R. H., Dawson, J., Johnston, M., Stewart, E., & Mattina, C. (2012b). Résilience, appartenance et tourisme au Nunatsiavut. *Études Inuits/Inuit Studies, 36*(2), 35–58.
Lemelin, R. H., Johnston, M., Dawson, D., Stewart, E., & Mattina, C. (2012c). From hunting and fishing to ecotourism: Examining the transitioning tourism industry in Nunavik. *The Polar Journal, 2*(1), 39–60.
Lemelin, R. H., Maher, P., & Liggett, D. (Eds.). (2013). *From talk to action: How tourism is changing the polar regions*. Thunder Bay: Centre for Northern Studies, Lakehead University.
Lequin, M., & Cloquet, I. (2006). Facteurs sociopolitiques influent sur la gouvernance de l'offre écotouristique : le cas des parcs nationaux au Nunavik. In C. Gagnon & S. Gagnon (Eds.), *L'écotourisme, entre l'arbre et l'écorce : De la conservation au développement viable des territoires* (pp. 230–266). Québec: Les Presses de l'Université du Québec.

Maher, P., & Lemelin, R. H. (2011). Northern exposure: Opportunities and challenges for tourism development in Torngat Mountains National Park, Labrador, Canada. *Polar Record, 47*(1), 40–45.

Martin, T. (2003). *De la banquise au congélateur. Mondialisation et culture au Nunavik*. Québec: Les Presses de l'Université Laval.

Martin, T. (2005). Modernité réflexive au Nunavik. *Globe, Revue internationale d'études québécoises, 8*(1), 175–206.

Martin, T. (2012). Parcs Nunavik : un outil pour inscrire la culture inuite dans le global. *Téoros, 31*, 13–18.

Milo, N., & Lariviere, V. (2012). La cogestion des parcs du nord du Québec en contexte d'incertitude. *Téoros, 31*(1), 82–91.

Ministère du Développement durable, de l'Environnement et des Parcs (MDDEP). (2007). *Plan directeur provisoire du parc national de la Kuururjuaq* (p. 48). Québec: Direction du Patrimoine écologique et des Parcs Service des Parcs, Québec.

Moore, S., & Weiler, B. (2009). Tourism-protected area partnerships: Stoking the fires of innovation. *Journal of Sustainable Tourism, 17*(2), 129–132.

Müller, D. K., & Jansson, B. (Eds.). (2007). *Tourism in peripheries*. New York: CABI.

Natcher, D., Felt, L., & Procter, A. (Eds.). (2012). *Settlement, subsistence and change among the Labrador Inuit* (p. 295). Winnipeg: University of Manitoba Press.

Parcs Canada. (2010). *Tongait KakKasuangita SilakKijapvinga, Parc national du Canada des Monts-Torngat, Plan directeur*, Ottawa, Gouvernement du Canada. Retrived from http://publications.gc.ca/collections/collection_2010/pc/R64-105-81-2010-fra.pdf

Plummer, R., & Armitage, D. (2007). A resilience-based framework for evaluating adaptive co-management: Linking ecology, economics and society in a complex world. *Ecological Economics, 61*(1), 62–74.

Rodon, T., & Grey, M. (2009). The long and winding road to self-government: The Nunavik and Nunatsiavut experience. In F. Abele, T. J. Courchene, F. L. Seidle, & F. St-Hilaire (Eds.), *Northern exposure: Peoples, powers, prospect in Canada's North* (pp. 317–345). Montréal: Institute for research on Public Policy.

Ross, A., Pickering Sherman, K., Snodgrass, J. G., Delcore, H. D., & Sherman, R. (2011). *Indigenous peoples and the collaborative stewardship of nature: Knowledge binds and institutional conflicts*. Walnut Creek: Left Coast Press.

Thomas, A. (2012). L'écotourisme au Nunvik: manifestation de la postermodernité. *Études Inuits/Inuit Studies, 36*(2), 79–98.

Vander Klippe, N. (2008, December). Labrador rising. *Up Here*, 61–65.

Winter, G., & Durhaime, G. (2002). Cooperative societies in Greenland and Nunavik: A lesson on the importance of supporting structures. *Journal of Rural Cooperation, 30*(1), 25–41.

Conceptual and Institutional Frameworks for Protected Areas, and the Status of Indigenous Involvement: Considerations for the Bering Strait Region of Alaska

Julie Raymond-Yakoubian

Abstract The Bering Strait region of Alaska is a culturally, economically, biologically, and politically important area of the Arctic. Like the rest of the arctic, this area is experiencing rapid and dramatic changes, both climate- and development-related. From the perspective of many indigenous residents, there is a growing need for protections – particularly in relation to the marine environment – in the Bering Strait region. This chapter reviews some of the existing protections that are in place and the status of indigenous involvement in them. The pressing need for additional protected areas is considered in light of the diverse issues and challenges facing the area such as commercial fishing, increasing marine traffic, climate change and resource development. I argue that it is critical to include indigenous residents of the region in the development, creation and maintenance of protected areas. I also argue that effective methods for protection can extend beyond typical western understandings of the nature, process and meaning of protection as defining an area where activities are allowed or prohibited.

Keywords Bering Sea • Alaska • Indigenous • Climate change • Vessel traffic • Fishing

1 Introduction

The western portion of Alaska is a remote and sparsely populated area in comparison to other regions of the United States. The Seward Peninsula region of Alaska is the westernmost portion of North America and is directly adjacent to the Bering Strait. Despite its comparative remoteness, the Bering Strait region is receiving increased, and in some cases unwanted, attention as an international strait at the center of expanding global marine commerce and resource development, as well as

J. Raymond-Yakoubian (✉)
Kawerak, Inc., P.O. Box 948, Nome, AK, USA
e-mail: JRaymond-Yakoubian@kawerak.org

T.M. Herrmann, T. Martin (eds.), *Indigenous Peoples' Governance of Land and Protected Territories in the Arctic*, DOI 10.1007/978-3-319-25035-9_5

experiencing an increase in climatic variability. These conditions have led to increased political attention being placed on the Bering Strait as a critical link in U.S. national security on a variety of levels, as well as being increasingly important to the security of global commerce (Clement et al. 2013). The results of changing climatic conditions, primarily reductions in sea ice extent and a longer ice-free season, have led to more vessels transiting through the Strait and more resource development activities such as offshore oil exploration. The total population of the region is approximately 9,900 people, about 75 % of whom are Alaska Native or Native American (U.S. Census Bureau 2013a). The region is the homeland of three distinct groups of indigenous peoples, Yup'ik, Inupiat and Saint Lawrence Island Yupik (collectively known as "Inuit"). There are 20 federally recognized tribes in the region. The indigenous and non-indigenous residents of the Bering Strait region live in 16 year-round occupied communities and utilize numerous other camps and locations seasonally (Fig. 1). The villages range in size from 120 residents in Little Diomede to 719 residents in Unalakleet. Nome, which is the largest and the "hub" community for the region, has a population of approximately 3700 (U.S. Census Bureau 2013b).

The marine environment of the Bering Strait and northern Bering Sea region is characterized by high productivity and is a biologically diverse ecosystem (e.g. Hunt and Stebeno 2002; Sigler et al. 2011; Grebmeier 2012). The region is a vital

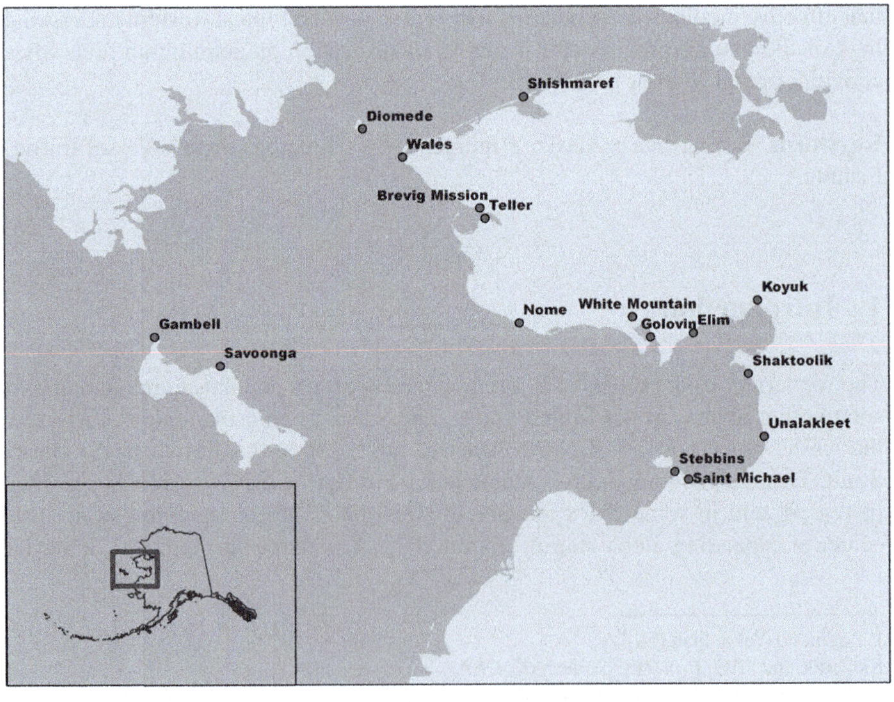

Fig. 1 The Bering Strait region of Alaska

area for marine mammals, sea birds and other marine species (e.g. Smith 2010; Garlich-Miller et al. 2011; Speer and Laughlin 2011; Laidre et al. 2008). The largest marine mammal migrations in the world pass through the region in the fall and spring (associated with ice formation and retreat). The waters of the Bering Strait cover an underwater "bridge" connecting the shallow continental shelves of the Bering and Chukchi Seas. Because of the diversity and productivity of the area, the Bering Strait and Saint Lawrence Island have both been identified as areas that meet and exceed the Convention on Biological Diversity's criteria for Ecologically and Biologically Significant Areas (EBSAs) and have been called "Super EBSAs" (CBD 2008; Speer and Laughlin 2011; Laughlin et al. 2012; McConnell et al. 2013). These characteristics of the region have also, of course, supported and shaped indigenous cultures for millennia.

Indigenous communities in the Bering Strait region continue to carry out subsistence hunting, fishing and gathering traditions in addition to participating in local cash economies. While no comprehensive statistics exist, the majority of households in the region (indigenous and non-indigenous alike) utilize at least a small amount of subsistence resources thought the year. Some indigenous households utilize thousands of pounds of subsistence foods per year, typically consuming such foods every day (Ahmasuk et al. 2008: 45). Marine mammals in particular are of special significance to the indigenous residents of the region, with some households harvesting over 2600 pounds of marine mammals in a year (Ahmasuk et al. 2008: 189) (see Fig. 2). There are specific traditions and practices related to the harvest and treatment of various animal and plant resources (e.g. Oquilluk 1973; Fienup-Riordan 1994; Kawerak 2013a, b, c). For example, some communities maintain

Fig. 2 Seal meat drying in a Bering Strait region community (J. Raymond-Yakoubian)

traditional beliefs regarding hunting, such as that hunting success is the result of animals giving themselves to hunters who are respectful towards animals (including using well-maintained gear, not being boastful, sharing meat, and other practices). These resources, and the practices and beliefs associated with them, are critical to local definitions of identity and are closely connected to individual and community well-being.

The maintenance of healthy, functioning ecosystems is a priority for the indigenous residents of the region (e.g. Gadamus 2013). Because many of the activities of subsistence practitioners are carried out in marine waters, the integrity of the ecosystem correlates with the ability of residents to continue to harvest resources and practice their subsistence way of life. Hunters and others travel in small boats in challenging conditions, often very far from shore. Hunters pursue very large mammals such as bowhead whales, as well as smaller animals like walruses and bearded seals (which can weigh more than 3500 and 500 pounds, respectively). Hunting activities can be extremely dangerous when all factors are considered (e.g. unpredictable weather, distance from shore, small size of boats, large size of animals, animal behavior, etc.). Region residents also fish in marine waters and gather resources such as seaweed, clams and other marine plant and invertebrate life. Small boats are also used to travel between communities and between communities and camps. Additional, or modified, protected areas, created through consultation and collaboration with indigenous people, may benefit the interests of indigenous communities, conservationists, researchers, and others.

Protected areas have been created in the marine waters of the northern Bering Sea and Bering Strait region. A "protected area" can be broadly defined as an area that has some sort of regulation on the type of activities that take place within its boundaries (this may be a time or space restriction or both). Below, some protected areas that currently exist and what they are meant to protect are described, some threats to the marine environment and communities that utilize it are outlined, and the ways that indigenous people have been and can be involved in the development of protected areas are presented. The varied forms that marine protections could take, and indigenous perceptions of protected areas, are also discussed.

2 Existing Protected Areas

The state of Alaska has jurisdiction over waters from 0 to 3 miles offshore and the federal government of the United States has jurisdiction over waters from 3 to 200 miles offshore. There are multiple protected areas in the marine waters off of Alaska (Witherell and Woodby 2005). The National Marine Fisheries Service (NMFS), North Pacific Fishery Management Council (NPFMC) and U.S. Fish and Wildlife Service (USFWS) are the federal agencies that are most active in the area. Each agency or body has different responsibilities and processes related to establishing a protected area and for consulting with tribes. Below is a discussion of some of the protected areas that exist in federal waters in the northern Bering Sea and Bering Strait region.

2.1 *Northern Bering Sea Research Area*

The Northern Bering Sea Research Area (NBSRA) encompasses a large portion of the northern Bering Sea, extending from approximately Saint Matthew Island north up to the middle of the Bering Strait (Fig. 3). The North Pacific Fishery Management Council (a management body authorized by the federal government) created the NBSRA to study the impacts of bottom trawl fishing on benthic and epibenthic fauna (NMFS 2011). Bottom trawl fishing is prohibited in the NBSRA until the NPFMC formally re-opens the area. To reopen the area the NPFMC would need to create a fishery management plan, receive input from its various committees, advisory panels and the public, and take a formal Council-member vote (NPFMC

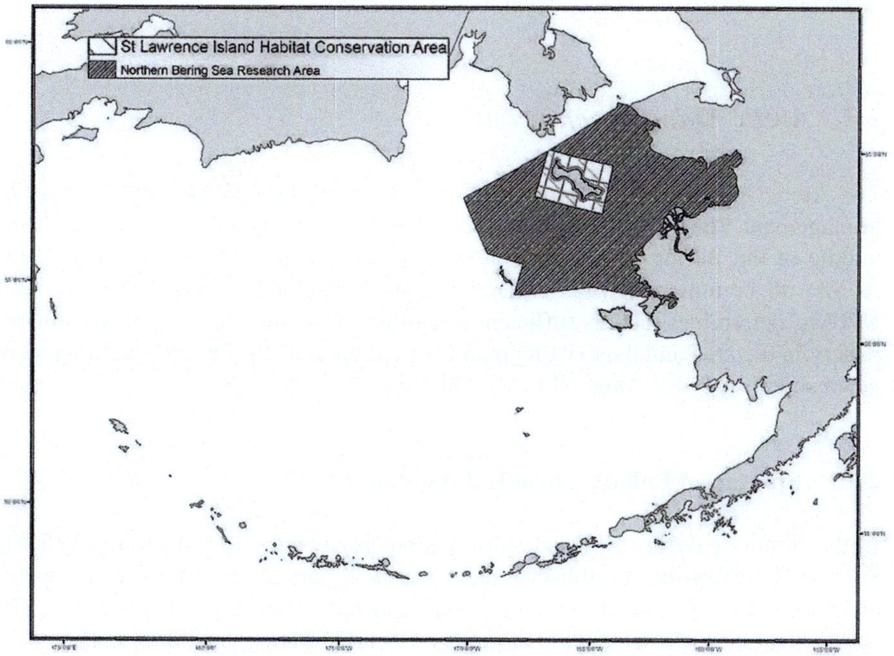

Fig. 3 Northern Bering Sea Research Area and Saint Lawrence Island Habitat Conservation Area (map from NMFS 2008:7)

2009a). Bottom-trawling is a very destructive fishing method that disturbs benthic habitats and fauna (NRC 2002; Stiles et al. 2010) and may cause prey depletions, which can create cascading impacts throughout the water column and entire ecosystem (e.g. Bluhm and Gradinger 2008). Many benthic and epibenthic fauna are important prey sources for other marine life.

2.2 *Saint Lawrence Island Habitat Conservation Area*

The Saint Lawrence Island Habitat Conservation Area (SLIHCA) is an area around Saint Lawrence Island where bottom trawl fishing is prohibited (Fig. 3). The NPFMC closed this area to conserve blue king crab habitat and to reduce conflict between subsistence fisheries and other activities (NMFS 2012). This area will remain closed to commercial bottom trawl fishing permanently, unless the NPFMC modifies the existing Fishery Management Plan for Groundfish of the Bering Sea and Aleutian Islands Management Area (*ibid.*). This area, along with the NBSRA, is not exempt from "research" activities approved by the NPFMC and research has taken place (see *Tribal Involvement*, below).

2.3 *Arctic Management Area*

The Arctic Management Area (AMA) was created through the Arctic Fisheries Management Plan (AFMP) and covers all federal waters in the Arctic from the middle of the Bering Strait north and east (Fig. 4). Through the authority of the AFMP, all commercial fisheries are currently prohibited in the AMA until the NPFMC determines there is sufficient scientific information to design a sustainable fishery in the area and the existing plan is modified to allow for the development of new fisheries (NMFS 2009; NPFMC 2009b).

2.3.1 Spectacled Eider Critical Habitat Areas

In the northern Bering Sea and Bering Strait region, the U.S. Fish and Wildlife Service (USFWS) has established several critical habitat areas for the spectacled eider in Norton Sound and south of Saint Lawrence Island (USFWS 2001). The spectacled eider is a sea duck that preys on benthic fauna while at sea. The entire world population of the species winters at a polynya south of Saint Lawrence Island. This critical habitat was established under the Endangered Species Act (ESA) and prohibits the destruction or adverse modification of the designated habitat, which has been deemed crucial to the survival of the species. Special consultations under the ESA are required for actions proposed within the critical habitat areas.

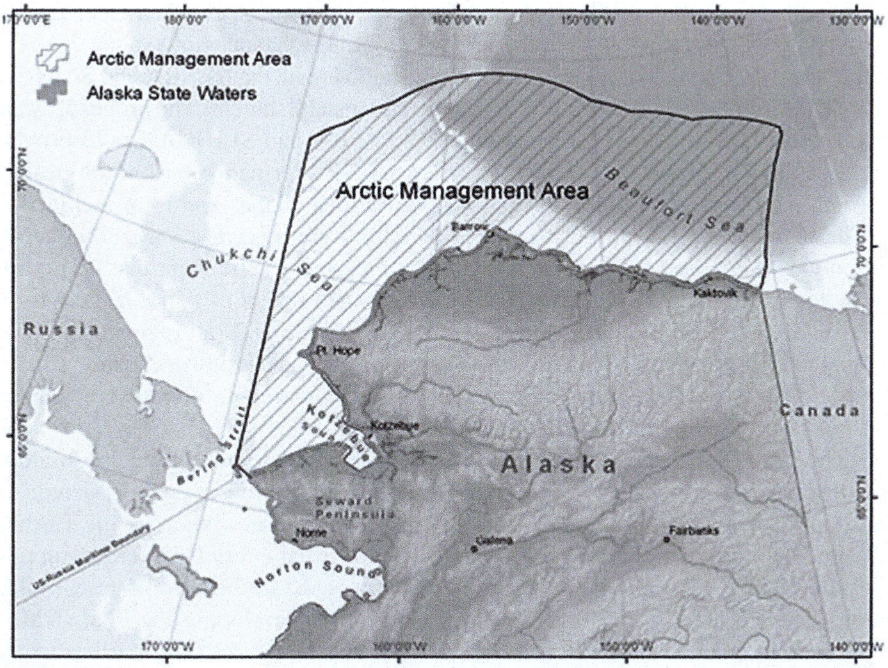

Fig. 4 Map of the area included under the Arctic Fishery Management Plan (from NPFMC 2009b:1)

2.3.2 Tribal Involvement

Some of the major protected areas in the northern Bering Sea and Bering Strait region were outlined above. The level of tribal involvement in the development, implementation and ongoing management of those areas has varied. As noted earlier, there are 20 federally recognized tribes in the Bering Strait region. Federal recognition gives tribes special status as "domestic dependent nations" with "inherent sovereign powers" (Executive Order 13175 [EO 13175]). This status entitles tribes to formal government-to-government consultation with the federal government regarding policies or actions that may impact tribal rights or resources. The State of Alaska, on the other hand, does not recognize tribes as having any special rights to consultation.

Through the consultation process tribes are given the voice they are entitled to under federal policy and regulations – if carried out in a timely and meaningful way. Unfortunately, the tribal consultation process does not always take place, and often when it does it is not conducted in a meaningful or timely fashion. No formal tribal consultation was carried out in conjunction with the establishment of the NBSRA or the SLIHCA, the AFMP, or during formulation of the spectacled eider critical habitat. Tribes were involved in the various processes as 'stakeholders'.[1]

During the process of reviewing the proposal to create the NBSRA and SLIHCA, tribal and community support was given for their establishment. The proposals were supported because it was believed that the NBSRA and SLIHCA would provide protections for subsistence. However, after the NBSRA had been formally established, a bottom trawl research plan was developed. Tribes were not included in discussions about the purpose or need for this trawl research and subsequently strenuously opposed research trawling and supported a permanent ban on bottom trawling in the Area (Bullard 2010). Despite tribal objections and a lack of tribal consultation, the research trawl took place in 2010. Bering Strait region tribes have remained deeply involved in trying to establish a permanent closure of the NBSRA to bottom trawl activities and have demanded tribal consultation on several aspects of this issue (Raymond-Yakoubian 2012).

The Arctic Fisheries Management Plan was created without formal consultation with tribes in the Bering Strait region. "Outreach"[2] activities were conducted; for example public presentations to tribally-authorized entities, such as the Eskimo Walrus Commission and Kawerak, Inc.[3] Overall, tribal entities expressed support for the Management Plan, though they also expressed a desire for formal consultations and a more meaningful role in the process of developing the Plan (NMFS 2009). If changes to the Management Plan are proposed in the future, tribes would request and expect formal consultation and an active role in the formulation of any new activities in the Arctic Management Area.

No formal tribal consultation was carried out during the designation of critical habitat for spectacled eiders. The USFWS held public meetings in various locales and conducted 'outreach' activities to indigenous communities and individuals. While the USFWS maintains that they have included tribal knowledge in their habitat designation process, tribes did not have the opportunity to formally participate in it through a consultation process (USFWS 2001).

In many cases, federal agencies only minimally engage in the tribal consultation process, which is currently the most effective process by which tribes can meaningfully engage in federal policy and rule making. Unfortunately, there is little account-

[1] The term 'stakeholder' generally refers to any member of the public or a group that is interested in a particular action. Tribes that are federally recognized have a higher status when it comes to federal actions that may impact them, and do not want to be treated as 'stakeholders'.

[2] The NPFMC has argued, and the Department of Commerce (DOC) recently codified, that they are not required to carry out formal tribal consultation activities (DOC 2013). The NPFMC conducts "outreach" activities, as described in the meeting notes from the first meeting of their "Rural Community Outreach Committee" (NPFMC 2009c). This has caused consternation amongst tribes because of the role that the NPFMC plays in decision making in federal marine waters (Raymond-Yakoubian 2012).

[3] Kawerak, Inc. (Kawerak) is an Alaska Native non-profit tribal consortium serving the 20 tribes of the Bering Strait region. Among other activities, Kawerak conducts independent research and advocates for policies and management actions that are supported by member tribes. For more information see www.kawerak.org

ability for federal agencies that do not carry out consultation appropriately. For example, some agencies will send a letter to all federally recognized tribes and tribal organizations in Alaska briefly outlining an upcoming activity, policy change or other matter; after sending the letter, little or no effort is made to determine which particular tribes may be most impacted or concerned about an action. Agencies often call such activities "tribal consultation," but it at most constitutes doing the bare minimum to claim this. Rather than a good faith effort to conduct meaningful consultation, these actions could be more appropriately characterized as 'going through the motions.' In these cases, if no responses are received from tribes they are then treated simply as 'stakeholders' in the process. Tribes are rarely asked to help develop or design policies or actions. Furthermore, in the case of interactions with the NPFMC, tribal input is often received after multiple detailed alternatives have been designed and the NPFMC, along with its parent agency NMFS, are at a point of being unwilling to consider redesign.

3 Challenges from the Perspective of Tribes

Bering Strait region tribes are faced with many threats to their subsistence way of life, especially threats to the marine resources and environment that they rely upon. As interest in the Arctic continues to grow, tribes struggle to maintain their involvement in the myriad issues, processes and policies that will directly impact them and future generations of Bering Strait residents. Tribes are very aware of the interconnections between various parts of the marine ecosystem and how impacts to one part of the food web may cascade through and impact other parts. The threats described below are of greatest concern to indigenous peoples in the Bering Strait region. This information has been obtained by the author during interactions with Bering Strait region tribes at public meetings, tribal consultations, Kawerak Board of Director's meetings, and other activities such as workshops and Kawerak Social Science Program research activities.

3.1 Bottom Trawl Fishing

Bottom trawl fishing activities can be extremely damaging to benthic environments (e.g. NRC 2002; Stiles et al. 2010) which are habitat for marine mammal and migratory bird prey species as well as for corals, sponges and other benthic life that support or provide habitat and prey for multiple marine species. Bering Strait region tribes are extremely concerned about the damage to benthic habitat that trawling may cause, as well as the impacts of that damage up the food chain. They are also worried about the added stress that would be placed on subsistence species if commercial fishing vessels and a bottom trawl fishery were active in the northern Bering Sea. The impacts of the actual fishing may also cause depletions in certain fish species which could also have cascading effects throughout the marine food chain (e.g. Garcia et al. 2003).

Bering Strait region tribes would like to see the entire northern Bering Sea remain closed to bottom trawl fishing. This desire for permanent protection is in contradiction to the desires of commercial fishers and the intent of the establishment of the NBSRA.

3.2 *Arctic Fishing*

Commercial fishing in the Arctic is currently prohibited under the Arctic Fisheries Management Plan (NMFS 2009; NPFMC 2009b). As discussed above, Bering Strait region tribes supported the creation of the AFMP and AMA by the NPFMC. Many tribes would prefer to keep Arctic waters off-limits to large-scale commercial fishing and supported the creation of this fisheries management plan. Tribes are concerned about the potential for expansion of fisheries northward as climate change modifies the distribution and range of various fish species (e.g. Meuter and Litzow 2008; Stram and Evans 2009; Cheung et al. 2010). The expansion of commercial fisheries northward would place additional pressures on marine subsistence species and subsistence practitioners.

3.3 *Salmon Bycatch*

The Bering Sea pollock fishery, which operates in federal waters, has caught very large amounts of primarily Chinook and chum salmon as bycatch in the past, and has the potential to do so each fishing season. These salmon are caught while large fishing vessels are trawling for pollock, and many originate in western Alaska (e.g. Guthrie et al. 2012; Kondzela et al. 2012). These salmon cannot be legally retained and the majority is thrown back into the ocean dead or dying. Salmon returns in the Bering Strait region (Seward Peninsula and Norton Sound areas) have been declining for several decades and many restrictions on subsistence salmon harvests have been enacted. Tribes believe that salmon bycatch in commercial fisheries is likely playing a role in these declines in salmon returns.

Various measures have been put in place by the NPFMC in an attempt to stem salmon bycatch in the pollock fishery. These measures have included time and area closures, including rotating hot spots, and salmon bycatch caps (NPFMC 2009d; NPFMC 2012). Tribes are extremely concerned about salmon bycatch in the pollock fishery and its impact on residents' ability to harvest salmon for subsistence. Tribes have argued for more and stronger measures surrounding the pollock fishery and have been involved in the development of some of these measures, but not to the degree that they have requested and wanted to be. Many of the alternatives that were eventually chosen for implementation were developed prior to any formal tribal consultation being conducted. Bering Strait region tribes are concerned that poor salmon returns have and will continue to contribute to loss of a nutritionally and economically important food, a culturally preferred food, and to the loss of cultural traditions and knowledge because they are not able to carry out traditional salmon fishing practices.

3.4 Increasing Vessel Traffic

Vessel traffic through the Bering Strait region is increasing at a steady rate and is predicted to continue to increase (Laughlin et al. 2012; McConnell et al. 2013; CMTS 2015). Tribes in the region have multiple concerns about existing ship traffic and the likelihood of it increasing. Some of the threats related to ship traffic that Bering Strait region tribes have identified include pollution (ship discharge into the water and air, spills) that will contaminate the marine environment and subsistence foods, noise disturbance negatively impacting marine mammals, shipping lanes disturbing marine mammal migrations, ship strikes to marine mammals, lack of spill response infrastructure in the region, subsistence hunter/small boat collisions with or interference from large ships, and lack of tribal participation in shipping-related policy development (e.g. Kawerak 2013d). The majority of decisions regarding vessel traffic through the strait are made in Washington, D.C. or in international contexts – typically far from the communities that are and will experience impacts from such traffic. As a result, Bering Strait region tribes have been very minimally involved in planning and policy development related to increasing vessel traffic. Tribal entities have submitted comments to various entities (e.g. Ray 2011, 2012; Kawerak 2013d; 2015; Raymond-Yakoubian 2013b) about their concerns and the need for shipping-related protections and have continually expressed their desire to be meaningfully involved in policy development and other processes surrounding this issue.

3.5 Resource Development

Offshore resource development activities are taking place in both the northern Bering Sea and north of the Bering Strait. Multiple corporations are pursing oil exploration and development activities in the Chukchi and Beaufort Seas. Tribes are concerned about oil spills, drilling-related ship traffic, noise from drilling-related rigs and ships disturbing marine mammals, icebreaker noise and disturbance of sea ice habitats, and lack of spill response infrastructure in the area (e.g. Kawerak 2013d). These concerns were exacerbated by the events of 2012 involving Shell Oil Company in which one ship dragged anchor and drifted while in harbor, their drilling operation in the Chukchi Sea was forced to shut down the day after it began because of sea ice, and a drilling rig broke free of its tow lines and ran aground

(Clement et al. 2013: 16, 19). Bering Strait region tribes have been minimally involved in the environmental and regulatory process surrounding oil exploration and development. Federal regulators and oil corporations have focused on engagement with tribes physically closer to the development activities, rather than on the extent of impacts to species connected to communities in the Bering Strait region.

In the Norton Sound area, in the vicinity of Nome, offshore gold dredging activities have increased over the past several years. These floating dredges use large suction hoses to suck up benthic sediments and then pass them through sluices to obtain gold. These are mostly small operations, but their number has increased and there is an interest by larger mining corporations in developing larger offshore operations (e.g. Jewett et al. 2013). Tribal concerns related to dredging include the impacts to salmon from the turbidity caused by dredging, the re-introduction of settled contaminants into the water column (such as mercury), and dredge interference with nearby subsistence salmon fishing activities (i.e. dredges becoming entangled in salmon nets) (e.g. Bullard 2012).

3.6 *Climate Change*

The Arctic is experiencing a variety of climate change-induced effects (ACIA 2005; NCADAC 2013; IPCC 2013). Residents of the northern Bering Sea and Bering Strait region have experienced changes to fisheries, to sea-ice, to storm severity, in wind patterns (i.e. strength, direction and seasonality), as well as to other aspects of the environment (e.g. Oozeva et al. 2004; Kwok et al. 2009; Krupnik et al. 2010; Raymond-Yakoubian 2009; 2013a; Raymond-Yakoubian et al. 2014). All of these changes, particularly changes to sea ice, have wide-ranging impacts on the indigenous residents of the region and on the threats they are experiencing, which were discussed above. These changes also create new challenges related to the gathering and processing of traditional food sources. Climate change impacts can magnify the effects of threats such as commercial shipping and resource development activities (e.g. Overland et al. 2011) by increasing the ice-free season in which activities can take place, for example.

4 Tribal Frameworks

There are four aspects of tribal engagements with protected areas that will be addressed here: (1) indigenous justifications for protections, (2) examples of existing or proposed protections which have, or would, involve indigenous participation, (3) conceptualizations of protections which have strong significance to tribal members but are not standard to western discourses on protection, and (4) the importance and neglect of relationship-based frameworks in western models of protection.

4.1 Indigenous Justifications

Tribal solutions to the threats and challenges described above may include a variety of protections and potential protected areas. Many Bering Strait tribes support the maintenance of existing protected areas and the creation of additional protective measures or protected areas. One challenge to developing new protections in the Bering Strait region (as well as other parts of the world) is determining why particular areas should be protected, as well as differing ideas of what "protections" should look like and consist of.

Bering Strait tribal communities are primarily concerned about the protection of subsistence-harvested resources, protection of the ecosystems those resources depend on, and the rights of tribal members to harvest those resources as their ancestors have done for generations. Some tribal solutions for protection of those rights and resources may appear similar to existing western models, but the reasons behind those solutions may often be different. For example, Bering Strait region tribes supported the creation of and support the continuing existence of, the NBRSA and the commercial bottom trawl prohibitions that go along with it. Tribes support this because of their concerns over disturbance to benthic habitats that are important for marine species and their own ability to carry out subsistence practices. The federal government, however, supported the creation of the area as a precautionary measure until it could conduct research to determine how much trawling was feasible in the area (e.g. as a potential precursor to trawling). So while tribes and the government both support the same overall protected area, tribes would like for it to exist in perpetuity, while it is the intent of the government to eventually allow bottom trawl fisheries into the area when feasible. Without proper tribal consultation (or without timely and meaningful consultation) there may be misunderstandings – by all parties – regarding the true intent and operation of various protections.

4.2 Indigenous Examples

Bering Strait tribes have proffered various ideas and suggestions as to the protections they would like to see implemented. Some have been implemented (primarily when they match western management and policy ideas) and some have not been implemented or investigated. A recent example of effective community-initiated work in the Bering Strait region comes from Saint Lawrence Island. The tribes of Saint Lawrence Island have created tribal ordinances to enforce proper walrus harvest practices, based on traditional Saint Lawrence Island Yupik values and practices (Metcalf and Robards 2011). These ordinances were developed by tribal members and the Eskimo Walrus Commission to ensure that walrus resources were protected and properly utilized. The USFWS, which is the federal manager of walrus, has agreed to these ordinances and allows tribes to locally enforce them.

Other tribal solutions to some of the current challenges in the region include placing some additional areas permanently off limits to benthos-disturbing activities and/or to vessel traffic (e.g. Kawerak 2013d). For example, tribes have discussed their desire to have protected areas around various islands in the northern Bering Sea, such as Little Diomede Island, which transiting vessels would be prohibited from entering. The purpose of such areas would be to reduce the potential for vessel disturbance to marine mammals and subsistence hunting activities, as well as reducing the potential for ship-island collisions. Additionally, certain times of the year are particularly important for marine mammal migrations, pupping and calving, and other activities. Tribes have also noted their desire to have certain areas in the northern Bering Sea and Bering Strait placed off limits to large vessel traffic during sensitive times of the year. This would likely involve season-specific routing measures for traffic.

Unfortunately, the reality is that we do not have a detailed idea of what 'tribal protections' can look like because the majority of protected areas and protections created in the region have all been imposed by outside entities and have been developed with a 'top-down' approach. There are abundant opportunities for managers and policymakers to collaborate on community-based and community-endorsed protections in the northern Bering Sea and Bering Strait region – protections that are based on both community and ecosystem needs as well as traditional knowledge and western science.

4.3 Alternate Conceptualizations of Protection

Indigenous attachments to place can be extremely strong and meaningful (e.g. Basso 1996; Thornton 2008). For Bering Strait tribal members, ties to place, including the vast expanse of the marine environment, are powerful and related to history, culture, tradition and identity. Relationships with places, like with other people and with animals, are related to cultural ideals of reciprocity and sharing. Indigenous residents of the Bering Strait region take the stewardship of lands, waters and resources very seriously.

One Bering Strait example of this comes from the village of Elim, in Norton Sound. During research with Elim residents on the topic of the importance of various subsistence resources, the author was told time and again about how one of the reasons people loved their community and the surrounding area so deeply was because they own the land (the village corporation is the landowner) and because of the access to subsistence resources that they depend on. Elim people take care of their land and waters because the land and waters take care of them. Ownership of, stewardship of, and a long relationship with, lands and waters, such as what indigenous residents of the Bering Strait have, are powerful means to obtain the protections of such places.

While the indigenous residents of Elim may own the land around them, because of various historical facts of law, indigenous people do not own or technically have any control over the marine waters of the northern Bering Sea and Bering Strait. They do, however, have millennia-long histories of utilizing and caring for these ecosystems and have longstanding relationships with the marine mammals and other marine resources that are a part of this environment. While some indigenous conceptualizations of importance may fall outside western norms of justifications to protect areas, indigenous residents of the region have significant reasons (nutritional, economic, cultural, spiritual) to be invested in the protection and care of marine waters.

4.4 The 'Relationship'

There is a rich body of literature documenting the importance of reciprocity and relationality to the nature of the interactions and interconnections between northern indigenous people and their environments (e.g. Hallowell 1960; Brightman 1973; Fienup-Riordan 1999; Tanner 1979; Berkes 1999; Cruikshank 2005; Willerslev 2007). This pattern holds true for Bering Strait peoples as well. However, protected areas and related management and policy actions in the Bering Strait, largely driven by non-local processes despite their substantial impact on local peoples, are largely devoid of any concept of the 'relationship'. In terms of human-human interactions, while a legal framework exists for respectful and meaningful relationship-based process of management and policy between indigenous residents on one hand and managers and policymakers on the other hand (the formal tribal consultation process as codified in EO 13175 and elsewhere), in practice this process is only rarely followed as intended. Additionally, in terms of conceptualizations of the environment, indigenous perspectives on the environment – which have long foregrounded concepts of reciprocity and relationality – are largely absent from western scientific, management, and policy discourse (though this has been changing to some degree in more recent times with the growing attention to Traditional Knowledge[4]). Instituting in practice a more relationship-based management and policy 'ecosystem' – in terms of both human-human and human-environment issues – would constitute a great step forward in the involvement of indigenous concerns relating to environmental protection. This would involve taking seriously both existing legal frameworks and human rights on the one hand and indigenous history and culture on the other; at their core both involve recognition of the basic existence and value of indigenous peoples and their knowledge.

In addition to specific protections to various resources and areas in the northern Bering Sea and Bering Strait area, tribes have a strong desire to be direct partici-

[4]Traditional Knowledge can be briefly defined as a holistic body of knowledge, held by a specific group of people, encompassing teachings, observations, experiments and experiences, and based on long-term and intimate contact with the local environment. This knowledge represents a way of life and often includes spiritual teachings, rules about proper behavior and resource use, and is passed on from generation to generation (see also Raymond-Yakoubian and Raymond-Yakoubian 2015).

Table 1 Examples of threats to subsistence and culture in the Bering Strait region, and potential solutions supported by tribes

Threats	Solutions
Bottom trawl fishing	Permanent closure of the Northern Bering Sea Research Area to commercial bottom trawl fishing
	No research trawl activities without prior, formal government-to-government tribal consultation
Arctic fishing	Permanent closure of the Arctic Management Area
Salmon bycatch	Very low hard-caps on salmon bycatch, which would result in immediate closures of the pollock fishery when met
Increasing Vessel Traffic	Strict regulation of ship discharge and noise
	Routing measures
	Creation of areas to be avoided permanently or seasonally
	Marine mammal observers on-board
	Sufficient response gear deployed in region
	Spill response and search and rescue training for village residents
	Free community access to ship location data (Automated Information System)
Resource exploration and development activities	Proven oil spill response plans
	Sufficient response gear deployed in region
	No discharge regulations for ships and drill rigs
	Limits on gold dredging activities at certain times of the year (related to salmon)
	Better oversight and enforcement of existing regulations related to gold dredging
Climate change	Implementation of the above solutions
	Global reductions in carbon emissions

pants in the design, justification and implementation of protections. While formal government-to-government tribal consultation is an important method through which to accomplish this, it is not the only way. Tribal experts have extensive and valuable information about Bering Sea and Bering Strait ecosystems, including animals, ocean currents, weather, and how to safely operate boats and equipment in Arctic waters. Agency and researcher partnerships with tribes and tribal entities are an additional way to ensure that tribes are meaningfully involved in region management and policy.

Following from the above, tribal consultation, or other partnerships that directly engage tribal members and their knowledge, have the potential to lead to community-initiated protections that both protect subsistence resources as well as meet the goals of federal, state and international managers and policymakers. Another benefit of such community-based work is the development of long-term relationships and the likelihood that that tribes and tribal members living and working in the region will support and endorse such measures because they had a voice and a hand in developing them (Table 1).

5 Conclusions

The Bering Strait region of Alaska is the home of three groups of indigenous people, all of whom rely on the marine environment to meet their subsistence needs and to fulfill diverse roles in their rich cultural and spiritual lives. Indigenous people are the original residents of the region and have longstanding knowledge of and relationships with the marine environment. This relationship is based on generations of observations, experiences, and knowledge-sharing between individuals, and is the basis for what is often called 'traditional ecological knowledge'.

Various protected areas currently exist in the marine environment of the northern Bering Sea and Bering Strait region. There are critical ecosystems and resources in the Bering Strait region, resources that are important to local tribes and communities, to the United States, and to the global community.

This chapter has argued that, from the perspective of indigenous people, additional protected areas or protections to specific resources are needed in the region. While protection of subsistence resources and a subsistence way of life is the priority for many indigenous residents, and western policymakers and resource managers may have different justifications for protections, this chapter suggests paths for jointly moving forward when engaging these issues. The majority of existing protections were put in place with little formal input from indigenous residents of the region and no formal tribal consultation. Indigenous residents are concerned about many threats to marine ecosystems and their subsistence way of life, have identified connections between various threats, and also have suggestions for solutions to these threats. Any new protected areas or protections, or any modifications to existing ones, must be developed in collaboration and consultation with the indigenous residents of the region. Western managers and policymakers must recognize the indigenous right to participate in policy and management decision making. If agencies and governments take the time to develop relationships with tribes and tribal members in the Bering Strait region it will not only enhance support for protections, but collaboration with tribes and their traditional knowledge base will lead to better decision making regarding the need for protected areas and determining what form protections could most effectively take.

Acknowledgements The author thanks Carolina Behe, Brenden Raymond-Yakoubian, Raychelle Daniel and Mary David for valuable suggestions and comments on earlier versions of this chapter and Lily Gadamus for creating Fig. 1. This chapter was made possible, in part, through funding from Kawerak, Inc., the Arctic Yukon Kuskokwim Sustainable Salmon Initiative (projects 601, 712, 1333), the U.S. Fish and Wildlife Service's Office of Subsistence Management (Study 10-151), the National Science Foundation (Grant No. ARC-1023686), the Oak Foundation (Grant No. 2010-OUSA-073), the National Fish and Wildlife Foundation (Grant No. 2010-0061-004, project funding provided by Shell), the North Pacific Research Board (project 1013 and a Graduate Student Award), EPSCoR Alaska (NSF award #OIA-1208927 and the state of Alaska) and the Norton Sound Economic Development Corporation, each of which has provided funding for various projects that have facilitated the author's involvement in and analysis of the issues discussed above. The author also thanks the tribes, communities and individuals in the Bering Strait region that have participated in research that this chapter is based on.

References

ACIA (Arctic Climate Impact Assessment). (2005). Cambridge: Cambridge University Press.

Ahmasuk, A., Trigg, E., Magdanz, J., & Robbins, B. (2008). *Bering Strait region local and traditional knowledge pilot project: A comprehensive subsistence use study of the Bering Strait region*. North Pacific Research Board Project Final Report Project #643. Nome: Kawerak, Inc.

Basso, K. (1996). *Wisdom sits in places: Landscape and language among the Western Apache*. Albuquerque: University of New Mexico Press.

Berkes, F. (1999). *Sacred ecology* (2nd ed.). New York: Routledge.

Bluhm, B., & Gradinger, R. (2008). Regional variability in food availability for Arctic marine mammals. *Ecological Applications, 18*(2), S77–S96.

Brightman, R. (1973). *Grateful prey: Rock Cree human-animal relationships*. Saskatchewan: Canadian Plains Research Center.

Bullard, L. (2010). Letter to Don Chapman dated November 15, 2010. Department of Commerce Tribal Consultation Concerns (letter includes multiple enclosures). On file with Kawerak, Inc., Nome, Alaska.

Bullard, L. (2012). Letter to Dan Sullivan, dated May, 15 2012. Off-shore mining concerns. On file with Kawerak, Inc. Nome, Alaska.

CBD (Convention on Biological Diversity). (2008, May 19–30). Decision IX/20: Marine and coastal biodiversity. In *Ninth meeting of the conference of the parties to the convention on biological diversity*, Bonn, Germany. Retrieved December 30, 2013, from http://www.cbd.int/decision/cop/default.shtml?id=11663

Cheung, W., Lam, V., Samiento, J., Kearney, K., Watson, R., Zeller, D., & Pauly, D. (2010). Large-scale redistribution of maximum fisheries catch potential in the global ocean under climate change. *Global Change Biology, 16*, 24–35.

Clement, J., Bengtson, J., & Kelly, B. (2013). Managing for the future in a rapidly changing Arctic. A report to the President. Interagency Working Group on Coordination of Domestic Energy Development and Permitting in Alaska (D. J. Hayes, Chair), Washington, DC.

CMTS (Committee on the Marine Transportation System). (2015). *International Council on Clean Transportation, 10-Year Projection of Maritime Activity in the U.S. Arctic Region*. Prepared for the U.S. Committee on the Marine Transportation System.

Cruikshank, J. (2005). *Do glaciers listen?: Local knowledge, colonial encounters, and social imagination*. Vancouver: University of British Columbia Press.

DOC (Department of Commerce). (2013, May 21). Tribal consultation and coordination policy of the U.S. Department of Commerce.

Executive Order 13175. (2000). Consultation and coordination with Indian Tribal Governments. *Federal Register, 65*(218), 67249–67252.

Fienup-Riordan, A. (1994). *Boundaries and passages: Rule and ritual in Yup'ik Eskimo oral tradition*. Norman: University of Oklahoma Press.

Fienup-Riordan, A. (1999). Yaqulget Qaillun Pilartat (What the birds do): Yup'ik Eskimo understandings of geese and those who study them. *Arctic, 52*(1), 1–22.

Gadamus, L. (2013). Linkages between human health and ocean health: a participatory climate change vulnerability assessment for marine mammal harvesters. *International Journal of Circumpolar Health, 72*, 207–215.

Garcia, S., Zerbi, A., Aliaume, C., Do Chi, T., & Lasserre, G. (2003). *The ecosystem approach to fisheries: Issues, terminology, principles, institutional foundations, implementation and outlook* (FAO fisheries technical paper number 443). Rome: Food and Agriculture Organization.

Garlich-Miller, J., MacCracken, J., Snyder, J., Meehan, R., Myers, M., Wilder, J., Lance, E., & Matz, A. (Eds.). (2011). Status review of the Pacific Walrus (*Odobenus rosmarus divergens*). Anchorage: U.S. Fish and Wildlife Service.

Grebmeier, J. (2012). Shifting patterns of life in the Pacific Arctic and Sub-Arctic Seas. *Annual Review of Marine Science, 4*, 63–78.

Guthrie III, C., Nguyen, H., & Guyon, J. (2012). *Genetic stock composition analysis of Chinook salmon bycatch samples from the 2010 Bering Sea trawl fisheries*. U.S. Department of Commerce, NOAA Technical Memorandum NMFS-AFSC-232.

Hallowell, A. (1960). Ojibwa ontology, Behavior, and world view. In S. Diamond (Ed.), *Culture in history: Essays in Honor of Paul Radin* (pp. 19–52). New York: Columbia University Press.

Hunt, G., & Stebeno, P. (2002). Climate change and the control of energy flow in the southeastern Bering Sea. *Progress in Oceanography, 55*, 5–22.

IPCC (Intergovernmental Panel on Climate Change). (2013). Climate change 2013: The physical science basis. In T. Stocker, D. Qin, G.-K. Plattner, M. Tignor, S. Allen, J. Boschung, A. Nauels, Y. Xia, V. Bex, & P. Midgley (Eds.), *Fifth assessment report of the intergovernmental panel on climate change*. New York: Cambridge University Press.

Jewett, S., Musgrave, D., Harper, S., & Gardner, L. (2013). *Nome offshore project 2012 environmental baseline studies* (Final report, Vol. 1). Chugiak: RWJ Consulting.

Kawerak Inc. (2013a). *Savoonga traditional knowledge of Ice Seals and Walruses*. Nome: Kawerak Social Science Program.

Kawerak Inc. (2013b). *Traditions of respect: Traditional knowledge from Kawerak's Ice Seal and Walrus project*. Nome: Kawerak Social Science Program.

Kawerak Inc. (2013c). *Seal and Walrus hunting safety: Traditional knowledge from Kawerak's Ice Seal and Walrus project*. Nome: Kawerak Social Science Program.

Kawerak Inc. (2013d). *Policy-based recommendations from Kawerak's Ice Seal and Walrus project*. Nome: Kawerak Social Science Program.

Kawerak Inc. (2015). *Port access route study: In the Chukchi Sea, Bering Strait, and Bering Sea PARS -USCG-2014-0941*. Letter to the U.S. Coast Guard dated June 1, 2015. On file with Kawerak Inc., Nome, Alaska.

Kondzela, C., McCraney, W., Nguyen, H., & Guyon, J. (2012). *Genetic stock composition analysis of chum salmon bycatch samples from the 2010 Bering Sea groundfish fisheries*. U.S. Department of Commerce, NOAA Technical Memorandum NMFS-AFSC-233.

Krupnik, I., Aporta, C., Gearheard, S., Laidler, G., & Kielsen Holm, L. (Eds.). (2010). *SIKU – Knowing our ice – Documenting Inuit sea ice knowledge and use*. New York: Spring.

Kwok, R., Cunningham, G., Wensnahan, M., Rigor, I., Zwally, H., & Yi, D. (2009). Thinning and volume loss of the Arctic Ocean sea ice cover: 2003–2008. *Journal of Geophysical Research*. doi:10.1029/2009JC005312.

Laidre, K., Stirling, I., Lowry, L., Wiig, O., Heide-Jorgensen, M., & Ferguson, S. (2008). Quantifying the sensitivity of Arctic marine mammals to climate-induced habitat change. *Ecological Applications, 18*(2), S97–S125.

Laughlin, T., Speer, L., & Brigham, L. (2012). Workshop report: Workshop to identify several viable options for the protection of Ecologically and Biologically Significant Areas (EBSAs) from the possible negative effects of shipping and other maritime activities in the Bering Strait Region. International Union for Conservation of Nature, Natural Resources Defense Council and the University of Alaska Fairbanks.

McConnell, M., Brigham, L., Laughlin, T., & Speer, L. (2013). Workshop II report: Workshop on expanded shipping and other marine activities and the ecology of the Bering Strait region. International Union for Conservation of Nature, Natural Resources Defense Council and the University of Alaska Fairbanks.

Metcalf, V., & Robards, M. (2011, September). *Tribal marine mammal ordinances of St. Lawrence Island*. Paper presented at the 27th Lowell Wakefield Fisheries Symposium, Anchorage.

Meuter, F., & Litzow, M. (2008). Sea ice retreat alters the biogeography of the Bering Sea continental shelf. *Ecological Adaptations, 18*(2), 309–320.

NCADAC (National Climate Assessment and Development Advisory Committee). (2013). Climate assessment report: Draft for public comment. January, 11 2013. Washington, DC: U.S. Global Change Research Program. Retrieved December 30, 2013, from http://ncadac.globalchange.gov/

NMFS (National Marine Fisheries Service). (2008). *Amendment 89 to the Fishery management plan for Groundfish of the Bering Sea and Aleutian Islands management area*. Retrieved December, 30 2013, from http://alaskafisheries.noaa.gov/sustainablefisheries/amds/89/language.pdf

NMFS (National Marine Fisheries Service). (2009). Environmental assessment/Regulatory impact review/Final regulatory flexibility analysis for the Arctic Fishery management plan and Amendment 29 to the Fishery management plan for Bering Sea/Aleutian Islands King and Tanner Crabs. Retrieved December 30, 2013, from https://alaskafisheries.noaa.gov/analyses/arctic/earirfrfa0809final.pdf

NMFS (National Marine Fisheries Service). (2011). *Northern Bering Sea research area webpage*. Retrieved July 31, 2012, from http://www.fakr.noaa.gov/npfmc/ruraloutreach/nbsra.html

NMFS (National Marine Fisheries Service). (2012). *Appendices: Fishery management plan for Groundfish of the Bering Sea and Aleutian Islands management area*. Retrieved December 30, 2013, from http://www.alaskafisheries.noaa.gov/npfmc/PDFdocuments/fmp/BSAI/BSAIfmpAPPENDIX.pdf

NPFMC (North Pacific Fishery Management Council). (2009a). *Navigating the North Pacific Council process* (2nd ed.). Anchorage: North Pacific Fishery Management Council.

NPFMC (North Pacific Fishery Management Council). (2009b). *Fishery management plan for fish resources of the Arctic management area*. Anchorage: North Pacific Fishery Management Council. Retrieved December 30, 2013, from http://www.npfmc.org/wpcontent/PDFdocuments/fmp/Arctic/ArcticFMP.pdf

NPFMC (North Pacific Fishery Management Council). (2009c). *Rural community outreach committee meeting report*. Anchorage: North Pacific Fishery Management Council. Retrieved August 12, 2009, from http://www.npfmc.org/wpcontent/PDFdocuments/rural_outreach/RCOCreport81209.pdf

NPFMC (North Pacific Fishery Management Council). (2009d). *Bering Sea Chinook salmon bycatch management*, Vol. I, Final environmental impact statement. Anchorage: North Pacific Fishery Management Council. Retrieved December 30, 2013, from http://www.alaskafisheries.noaa.gov/sustainablefisheries/bycatch/salmon/chinook/feis/eis_1209.pdf

NPFMC (North Pacific Fishery Management Council). (2012). *Bering Sea Non-Chinook salmon PSC management measures, initial review draft environmental assessment*. Anchorage: North Pacific Fishery Management Council. Retrieved December 30, 2013, from http://www.npfmc.org/wp-content/PDFdocuments/bycatch/ChumEA_InitRev_312.pdf

NRC (National Research Council). (2002). *Effects of trawling and dredging on seafloor habitat*. Washington, D.C.: National Academy Press.

Oozeva, C., Noongwook, C., Noongwook, G., Alowa C., & Krupnik, I. (2004). *Watching ice and weather our way: Sikumengllu Eslamengllu Esghapalleghput*. Washington, DC/Savoonga: Arctic Studies Center, Smithsonian Institution/Savoonga Whaling Captains Association.

Oquilluk, W. (1973). *People of Kauwerak: Legends of the Northern Eskimo*. Anchorage: Alaska Methodist University.

Overland, J., Wood, K., & Wang, M. (2011). Warm Arctic—cold continents: Impacts of the newly open Arctic Sea. *Polar Research, 30*, 15787–15790.

Ray, L. (2011). Letter to Lieutenant Faith Reynolds, U.S. Coast Guard, May 2, 2011. Port Access Routing Study Comments. On file with the Kawerak Social Science Program, Nome, Alaska.

Ray, L. (2012). Letter to Commander Houck, U.S. Coast Guard, dated November 20, 2012. Comments to USCG PARS from Kawerak's Ice Seal and Walrus project. On file with the Kawerak Social Science Program, Nome, Alaska.

Raymond-Yakoubian, B., & Raymond-Yakoubian, J. (2015). *"Always taught not to waste": Traditional knowledge and Norton sound/Bering Strait Salmon Populations. Arctic-Yukon-Kuskokwim Sustainable Salmon Initiative Project 1333 Final Product*. Nome: Kawerak, Incorporated.

Raymond-Yakoubian, B., Kaplan, L., Topkok, M., & Raymond-Yakoubian, J. (2014). *"The world has changed": Iŋalit traditional knowledge of Walrus in the Bering Strait*. Final report to the North Pacific research board for project 1013. Nome: Kawerak Social Science Program.

Raymond-Yakoubian, J. (2009). *Climate-ocean effects on Chinook salmon: Local traditional knowledge component*. Nome: Kawerak Social Science Program.

Raymond-Yakoubian, J. (2012). Participation and resistance: Tribal involvement in Bering Sea fisheries management and policy. In C. Carothers, K. Criddle, C. Chambers, P. Cullenberg, J. Fall, A. Himes-Cornell, J. Johnsen, N. Kimball, C. Menzies, & E. Springer (Eds.), *Fishing people of the North: Cultures, economies, and management responding to change* (pp. 117–130). Fairbanks: Alaska Sea Grant and University of Alaska Fairbanks.

Raymond-Yakoubian, J. (2013a). When the fish come, we go fishing: Local ecological knowledge of Non-Salmon fish used for subsistence in the Bering Strait region. Final report (Study No. 10-151) to the U.S. Fish and Wildlife Service, Office of Subsistence Management, Fisheries Resource Monitoring Program. Nome: Kawerak Social Science Program.

Raymond-Yakoubian, J. (2013b, February 5–6). Bering Strait maritime symposium field notes, Nome, Alaska. On file with the author.

Sigler, M., Renner, M., Danielson, S., Eisner, L., Lauth, R., Juletz, K., & Logerwell, E. (2011). Fluxes, fins, and feather: Relationships among the Bering, Chukchi, and Beaufort Seas in a time of climate change. *Oceanography, 24*, 250–265.

Smith, M. (2010). *Arctic marine synthesis: Atlas of the Chuckchi and Beaufort Seas*. Anchorage: Audubon Alaska and Oceana.

Speer, L., & Laughlin, T. (2011). Workshop report: Workshop to identify areas of ecological and biological significance or vulnerability in the Arctic marine environment. International Union for Conservation of Nature and Natural Resources Defense Council. Retrieved December 30, 2013, from http://data.iucn.org/dbtw-wpd/edocs/Rep-2011-001.pdf

Stiles, M., Stockbridge, J., Lande, M., & Hirshfield, M. (2010). *Impacts of bottom trawling on fisheries, Tourism, and the marine environment*. Belize City: Oceana.

Stram, D., & Evans, D. (2009). Fishery management responses to climate change in the North Pacific. *ICES Journal of Marine Science, 66*, 1633–1639.

Tanner, A. (1979). *Bringing home animals: Religious ideology and mode of production of the Mistassini Cree Hunters*. New York: St. Martin's Press.

Thornton, T. (2008). *Being and place among the Tlingit*. Seattle: University of Washington Press.

U.S. Census Bureau. (2013a). State and County quick facts, Nome Census Area, Alaska. Retrieved November 1, 2013, from http://quickfacts.census.gov/qfd/states/02/02180.html

U.S. Census Bureau. (2013b). Population estimates: Places in Alaska listed alphabetically. Retrieved November 1, 2013, from http://quickfacts.census.gov/qfd/states/02/02180lk.html

USFWS (U.S. Fish and Wildlife Service). (2001). Endangered and threatened wildlife and plants; Final determination of critical habitat for the spectacled eider. *Federal Register, 66*(25), 9146–9185.

Willerslev, R. (2007). *Soul hunters: Hunting, animism, and personhood among Siberian Yukaghirs*. Los Angeles: University of California Press.

Witherell, D., & Woodby, D. (2005). Application of marine protected areas for sustainable production and marine biodiversity off Alaska. *Marine Fisheries Review, 67*(1), 1–27.

Part III
Cultural, Spiritual, Ecological, Aesthetic, Recreational and Economic Values of Protected Areas and Indigenous Lands and Territories in the Arctic

Protecting the 'Caribou Heaven': A Sacred Site of the Naskapi and Protected Area Establishment in Nunavik, Canada

John Mameamskum, Thora Martina Herrmann, and Blanka Füleki

Abstract Sacred Natural Sites play an essential role in the expression and transmission of culture, in the conservation of biodiversity, and are a vital means for the manifestation of cultural and spiritual values related to nature. In Nunavik, the Government of Québec, in partnership with the Kativik Regional Government recently created the Kuururjuaq National Park on 4,274 km^2 of tundra. A cultural important site for the Naskapi First Nation, the Caribou Heaven is situated within the limits of this new protected area. This chapter first provides an overview of the linkages between Aboriginal peoples and protected areas in Canada. It then illustrates the crucial role played by the caribou (Rangifer tarandus) in the socio-cultural, spiritual, and economic life of the Naskapi First Nation. Next, it explains how the ecological knowledge of the Naskapi was used to designate this culturally important place as an area of maximum protection, in order to ensure its protection and integrity. It finally describes how cultural and spiritual values, have formed the basis of co-management models of nature conservation in this park. The initiative is among the first of such efforts by the Government of Québec to give expression to the importance of and to provide protection to the sacred sites of First Nations.

Keywords Sacred sites • First Nations • Management • National parks • Sub-Arctic

J. Mameamskum
Naskapi Nation of Kawawachikamach, Schefferville, QC, Canada
e-mail: jmameamskum@naskapi.ca

T.M. Herrmann (✉)
Département de géographie, Université de Montréal, Montréal, QC H3C 3 J7, Canada
e-mail: thora.martina.herrmann@umontreal.ca

B. Füleki
International Advisor, Helvetas Swiss Intercooperation, Zurich, Switzerland
e-mail: pitypang.virag@gmail.com

T.M. Herrmann, T. Martin (eds.), *Indigenous Peoples' Governance of Land and Protected Territories in the Arctic*, DOI 10.1007/978-3-319-25035-9_6

1 Introduction

One of the most prominent forms of culture-based nature conservation has been the identification and protection of sacred natural sites, which often protect key ecosystems. Rural and Aboriginal societies, with their respective visions of the world, created protected areas well before the existence of Yellowstone National Park, the model which is the basis for current legislation, policy and management for the world's protected areas (Wild and McLeod 2008).

This chapter will highlight the spiritual dimension of protected areas, which is often neglected and under-appreciated in discussions on protected areas. We will explore the process of recognition and integration of a site that is culturally important and sacred to the Naskapi Nation, the Caribou Heaven, located at the heart of the new Kuururjuaq National Park in Nunavik, in order to ensure its protection and integrity. This initiative is one of the first in a series of measures by the Quebec government to convey the importance of Aboriginal nations' sacred sites and to ensure their protection. Several basic issues concerning sacred natural sites and protected areas will be dealt with in this chapter: how could the sacredness of a site contribute to conserving the biodiversity of fragile species? How should the seizing of public spaces for the development of national park projects incorporate this sacred dimension which connects First Nations to their natural environment? What status can be granted to this notion of space in the creation of a national park? This chapter is a call to managers of protected areas and to administrators and decision-makers to recognize spiritual values and sacred natural sites and their potential in contributing to the efficient management of protected areas.

2 Sacred Natural Sites and Protected Areas

The notion of sacred usually refers to a particular status given to objects, places or abstract characters in a given society. Sacred natural sites (forests, lakes, mountains, etc.) are natural areas that have a particular spiritual importance for a people, a society or a community (Wild and McLeod 2008). They are places where the tangible meets the intangible. They are the expression of world views in which nature is a living being. Natural sacred sites include natural regions that are the embodiment of cultural, spiritual or religious values and that are hence recognized as being sacred by local and Indigenous communities, as well as natural areas recognized by institutionalized religions or faiths as being places of worship. Numerous sacred natural sites are areas of great importance for biodiversity conservation (Mallarach 2009). They are recognized as being the oldest protected areas of the world (Dudley et al. 2005).

Approximately 80 % of our planet's biodiversity and 95 % of the world's cultural diversity are found on land belonging to Indigenous and local communities (Sobrevila 2008). A large portion of these lands is recognized as sacred or contain sacred natural sites. Sacred natural sites are present on all continents and in nearly every country, in all types of protected areas and in all habitats (Brown et al. 2005; Mallarach and Papayannis 2007): sacred forests in India (Ormsby and Bhagwat 2010) and in Africa (Kouami et al. 2005); Madagascar's springs inhabited by the ancestors' spirits; the summit of Mount Kenya, the earthly resting place for Ngai (God) for the Kikuyus; the Kilauea volcano in which Native Hawaiians see the body of the volcano goddess Pélé (Bernbaum 2002); the sacred Araucaria araucana forest and tree of the Mapuche people in Chile and Argentina (Herrmann 2006); Mount Kailash, which, for Tibetan Buddhists, is the home of Demchok, the Buddha of supreme bliss; le Kakadu National Park, home of the ancestors of Australian Aboriginal peoples (Gillespie 1983); the San Francisco Peaks in Arizona (U.S.A.) which provide medicinal herbs and other benefits, such as water, health and well-being, to the Navajo and Hopi peoples (Bernbaum 2002); or the sacred mountains of Sinai and Moriah in the Judeo-Christian tradition. Sacred natural sites can be the homes of gods, spirits or ancestors, or are associated with hermits, prophets or saints. They can be places of ceremony or contemplation, of prayer and meditation. These few examples show the many forms that the spiritual dimension of protected areas can assume. Higgins-Zogib concludes that: "[...] millions of people have a special regard for and relationship with hundreds, or thousands, of protected areas not because of their importance to biodiversity but because of their spiritual values". (Higgins-Zogib 2008:50) In sum, sacred natural sites protect a great variety of habitats, and preserve traditional customs, practices and knowledge related to biodiversity conservation (Berkes 2008).

Sacred natural sites are also an integral part of cultural identity and play a key role in cultures and ways of life. For example, the Maoris in New Zealand see mountains as the frozen bodies of their ancestors. They symbolize their ethnic and personal identity. At tribal meetings, the Maoris introduce themselves first by saying the name of their tribe's mountain, then their lake or their waterway, and finally their chief (Bernbaum 2002). In many societies, sacred natural sites fulfill roles similar to those of protected areas established by law. Because of the spiritual values attributed to these sites, access and use restrictions often apply; for example, there can be spatial or temporal prohibitions, or closure of access to particular species in certain areas or during certain determined periods (generally grouped under the term taboo) (Colding and Folke 2001; Borrini-Feyerabend et al. 2010). Thanks to these management tools based on customary rules governing access, use, protection and restoration, disruption by humans is reduced, if not completely prevented. Thus, many sites remain in a natural or quasi-natural state. In other sacred natural sites, human influence is greater, but it is still possible to maintain a high level of biodiversity. This is notably the case with Globally Important Agricultural Heritage Systems (e.g., le *Parque de la Papa* en Peru) (Koohafkan and Boerma 2006; Amend et al. 2008; Argumedo 2008).

2.1 Cultural and Spiritual Values Associated with Biodiversity

An important issue in this context is that of the interactions between biodiversity and cultural diversity. Throughout the world, Indigenous peoples and local communities have a close relationship with biological diversity and often have an intimate understanding of their environment and its ecology. They know how to exploit numerous animal and plant species in many ways, transforming them into food, medicine or dyes, and they have developed cultural techniques for many useful plants (Nakashima and Roué 2002). This knowledge and know-how that have been amassed through the ages are collectively owned by the communities and are handed down orally from one generation to the next, notably in the form of cultural values and practices (myths, songs, dances, etc.) (Descola and Palsson 1996; Ellen and Fukui 1996; Turner et al. 2000; Maffi 2005). Human intervention on the environment, including managing it, is a social act that is largely determined by the cultural and spiritual values associated with biodiversity; its condition, its evolution and the services that it brings influence the cultural expression of peoples (Posey 1999). Many cultural practices are, in their existence and expression, the result of biodiversity (UNESCO and UNEP 2003). Cultural diversity, with its wealth of knowledge, know-how, beliefs, values and forms of social organization, and biological diversity, are profoundly interdependent, mutually reinforcing and contribute to the preservation of our planet. The concept of bio-cultural diversity, which has recently emerged, allows us to understand the complex bonds that exist between culture and nature (Maffi and Woodley 2010).

Sacred natural sites represent important reservoirs of this bio-cultural diversity and constitute a diversified set of examples of the relationship of man with nature, many of which originate in time immemorial. They make it possible to reveal the processes through which cultural beliefs and practices (myths, songs, stories, dances) create tight bonds between societies and nature, and thus uncover new approaches in the search for strategies and tools for preserving bio-cultural diversity. However, their contributions to the preservation of bio-cultural diversity have often been ignored or undervalued by state organizations charged with conservation, as well as by policies and laws.

2.2 Threatened Sacred Natural Sites

In certain cases, sacred natural sites belong to areas designated as protected by law, but they can also be located outside of official protected areas. In other cases, the protected areas cover a small part of a larger sacred landscape. In all these situations, the sacred natural sites present specific challenges in terms of their recognition, conservation and management. Numerous protected areas have been superimposed on traditional use areas of local or Indigenous communities. When establishing protected areas, the values and the importance of the sacred sites and their traditional uses were often ignored, which affects the fundamental rights of local and Indigenous communities (Verschuuren et al. 2010). This situation, which reflects the confrontation of two world views, has often led to conflict, which, in turn, has prevented cooperation between Aboriginal peoples or local communities and conservation institutions (Higgins-Zogib 2007). Other sacred natural sites have faced threats from outside, such as tourism, industrialization and urbanization (Oviedo and Jeanrenaud 2007).

2.3 *Recognition of Sacred Sites in International Legal and Political Frameworks*

In recent years, the importance of the contribution of sacred natural sites to the protection of the environment has increasingly been recognized, which has sparked growing interest in these sites as biodiversity conservation tools. Over the last few years, a certain number of international agreements have been signed which pertain in different ways to the cultural and spiritual values of protected areas. In 2004, the Convention on Biological Diversity (CBD) recognized the importance of sacred sites and defined the *Akwé: Kon Guidelines* for conducting studies on cultural impacts (SCBD 2004). Articles 8(*j*) and 10(*c*) of the CDB (ratified by Canada in 1992) also offer possibilities for action to protect sacred sites by societies that recognize them as such.[1] In addition, Article 7.1 of Convention No. 169concerning Indigenous and tribal peoples of the International Labor Organization (ILO 169) stipulates that these peoples:

> [...] shall have the right to decide their own priorities for the process of development as it affects their lives, beliefs, institutions and spiritual well-being and the lands they occupy or otherwise use, and to exercise control, to the extent possible, over their own economic, social and cultural development. In addition, they shall participate in the formulation, implementation and evaluation of plans and programmes for national and regional development which may affect them directly.[2]

In 2007, the United Nations declaration on the Rights of Indigenous Peoples (adopted by Canada in 2010) in Article 12, ensures that indigenous peoples have the right to practice and teach their spiritual rites and to maintain and protect their religious sites.[3] UNESCO's Convention for Safeguarding Intangible Cultural Heritage of 2003 aims to ensure "respect for the intangible cultural heritage of the communities, groups and individuals concerned" and to "raise awareness at the local, national and international levels of the importance of the intangible cultural heritage, and of

---

[1] Convention on Biological Diversity (CBD). Retrieved, July 25, 2014, from http://www.cbd.int/convention/text/default.shtml

[2] ILO Convention 169. Retrieved, July 25, 2014, from http://www.ilo.org/dyn/normlex/en/f?p=NORMLEXPUB:12100:0::NO::P12100_ILO_CODE:C169

[3] United Nations declaration on the Rights of Indigenous Peoples. Retrieved, July 25, 2014, from http://www.un.org/esa/socdev/unpfii/documents/DRIPS_en.pdf

ensuring mutual appreciation thereof" (UNESCO 2003). In 2003, the Fifth Annual World Parks Congress of the IUCN (International Union for Conservation of Nature) held in Durban gave great importance to the issue of sacred natural sites and produced for this purpose recommendations and policies such as the "Delos initiative on sacred natural sites in technologically developed countries" (DELOS 2014). According to Recommendation No. 13 of the Fifth Annual World Parks Congress of the IUCN, sites with cultural value can be designated as "sacred sites" when protected areas are established (IUCN 2003). Areas designated as sacred enjoy special protection.

Moreover, the Biosphere reserves,[4] the Ramsar sites,[5] the World Heritage sites,[6]Indigenous peoples' and community conserved territories and areas (ICCAs),[7] Globally Important Agricultural Heritage systems (GIAHS)[8] comprise a large number of cultural landscapes and sacred sites worldwide. These various international instruments (legal and political) provide ways of enabling the recognition and protection of sacred natural sites at the international and national levels (Gomez et al. 2010). This article provides an example of the application of Recommendation No. 13 of the IUCN in Northern Quebec, in Nunavik, where one of the first initiatives involving collaborative planning and management of a protected area is presently under way.

3 Indigenous People and Protected Areas in Canada

In 1975, the Crees of James Bay, the Inuits of Quebec and the governments of Quebec and Canada signed the James Bay and Northern. Quebec Agreement (JBNQA). Following that, on January 31, 1978, the Naskapis signed the Northeastern Quebec Agreement (NEQA). These agreements – the first modern treaties settling land claims – grant Aboriginal peoples, among other things, the right to participate fully in the management of parks and protected areas and ensure that they have the right to review future development projects in the entire portion of the territory governed by the agreement. The JBNQA notably lead to the establishment of a joint Committee on hunting, fishing and trapping – the first of what today are called "co-management organizations" – between provincial and federal governments and Cree and Inuit representatives on environmental issues. It allows the sharing of

---

[4] Man and Biosphere Programme (MAB) of UNESCO. Retrieved, July 25, 2014, from http://www.unesco.org/new/en/natural-sciences/environment/ecological-sciences/man-and-biosphere-programme/

[5] Ramsar Sites. Retrieved, July 25, 2014, from http://ramsar.wetlands.org/

[6] List of World Heritage sites. Retrieved, July 25, 2014, from http://whc.unesco.org/en/list/

[7] See ICCA Consortium. Retrieved, July 25, 2014, from www.iccaconsortium.org/ and also the ICCA registry UNEP-WCMC. Retrieved, July 25, 2014, from www.iccaregistry.org/

[8] Globally Important Agricultural Heritage systems (GIAHS). Retrieved, July 25, 2014, from http://www.fao.org/giahs/en/

responsibilities at many levels; this committee can notably advise the ministries concerned as to the establishment of new parks or new protected areas (CPC 2008). In 1984, the Inuvialuit Final Agreement (IFA) established management organizations similar to those of James Bay, in national and territorial parks. The largest protected area that falls within several jurisdictions is the Thelon Game Sanctuary which is co-managed by the governments of Nunavut and the Northwest Territories, Aboriginal organizations and by the Ministry of Indian Affairs and Northern Development.

Through these agreements, as well as others that are cooperative in nature, Aboriginal peoples have been involved in the establishment of more than a quarter of the surface area of the protected areas administered by federal, provincial and territorial governments. Two protected areas of over 10 km^2 were created in the Northwest Territories through Aboriginal land claim agreements and they are managed directly by them. Taken together, these agreements, entered into in Quebec and other regions, have changed how parks and protected areas are planned, established and managed in Canada. They have made it possible to include traditional Aboriginal knowledge in decision-making processes, to provide new economic possibilities for the Aboriginal peoples associated with the parks, and to give them an important role in managing the parks and conservation, both now and in the future.

4 Kuururjuaq National Park

Created in May 2009 by the government of Quebec in partnership with the Kativik Regional Government (made up of 14 northern villages, including the Naskapi Nation of Kawawachikamach, located directly below the 55th parallel) Kuururjuaq National Park[9] is the second park created in Nunavik. This park is part of the Quebec government's project to create a network of protected areas (corresponding to IUCN categories I, II, III, IV, V and VI) representing 12 % of its land and covering all biodiversity regions.[10]

With a total surface area of 4,461 km^2, the park covers nearly the entire watershed of the Koroc River and five other secondary watersheds. Incised in a valley, the Koroc River originates in the Torngat Mountains and flows 160 km before emptying into Ungava Bay. The Torngat Mountain range forms the eastern boundary of Kuururjuaq National Park . This rugged mountain range is the highest in eastern continental Canada (1,646 m in altitude) (Fig. 1).

For several millennia, groups of people have used the Koroc River, and the existence of a few archeological sites along its route indicate that the area was used by

---

[9] *Kuururjuaq* means "narrow valley" in Inuktitut.

[10] Québec covers 1,667,441 km^2. The Québec network of protected areas currently covers 9.11 % of the province. See also: Ministère du Développement durable, Environnement et Lutte contre les changements climatiques, *Pourcentage du territoire québécois en aires protégées*. Retrieved, July 25, 2014, from www.mddelcc.gouv.qc.ca/biodiversite/aires_protegees/registre/index.htm

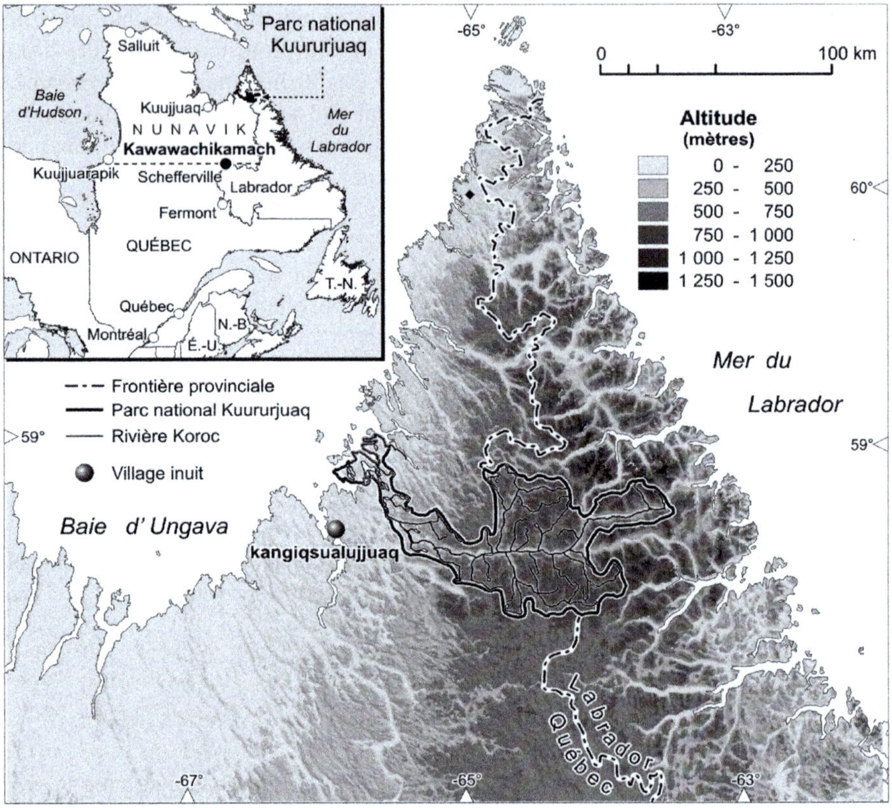

Fig. 1 Kuururjuaq National Park (Carthography: Marc Girard, Département de géographie, Université de Montréal, 2011)

Paleo-Eskimo groups traveling between the shores of Labrador and Ungava Bay (ARK 2005).

As for its natural heritage, because of its northern latitude, Kuururjuaq Park's land is made up of different ecosystems (i.e., the coastal environment, the forested area in the Koroc River valley and the tundra) and different habitats (e.g., aquatic habitats, wetlands, etc.). The park also contains more than 850 rare vascular plant species and 80 rare bryophyte species (Rousseau 1953; Ouellet 1978; Desponds 2004; Dignard 2004). In addition, the Koroc River valley contains a forest composed of black spruce and larch (a boreal enclave within an arctic environment), and Quebec's northernmost white birch stand. Some 126 bird species, 24 fish species and 39 mammal species, including the caribou, can be found within the territory of Kuururjuaq Park (ARK 2005).

There are two large herds of migratory caribou (*Rangifer tarandus*) in Québec: the Rivière-George (RG) herd and the Rivière-aux-Feuilles (RAF) herd (Figs. 2 and 3). These two migratory caribou herds which until recently were considered very

Fig. 2 Caribous (*Rangifer tarandus*) from the Rivière-George herd crossing the George River (© Photo: Thora Herrmann, August 2012)

healthy with an estimated population of 1,013,000 animals in 2001 (Courtois et al. 2003), have suffered a dramatic decline. Thus, the RG herd reduced from about 800,000 heads in 1993 to 14,200 in summer 2014 (Porter 2011; CBC 2014). The caribou of the RG herd are present in the Kuururjuaq Park area from May to October (Boudreau et al 2003; Jean and Lamontagne 2004). The territory of the Park is also home to a population of caribou of the mountain ecotype, represented by the Torngat Mountain herd. To date, very little is known about the latter. The caribou is central to the ecology of arctic environments. For the Cree, Inuit and Naskapi peoples living in the subarctic and arctic areas, the caribou is at the heart of their culture, their social life and their subsistence.

5 The Naskapi Nation and the Connections between Nature and Culture

The Naskapi are one of Quebec's ten First Nations. The majority of the 1,028 members of the Naskapi Nation live in the village of Kawawachikamach near Schefferville in northeastern Quebec (Fig. 1). Naskapi is the main language and is spoken by all the inhabitants, including at work. Tradition is still very much alive for the Naskapi

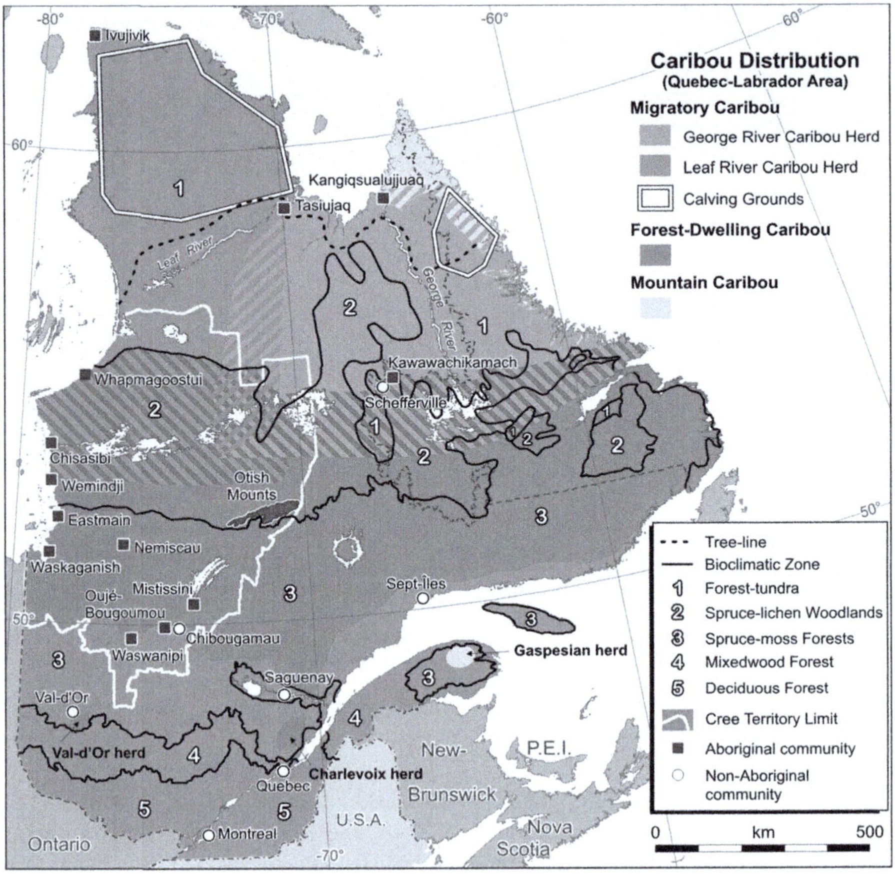

Fig. 3 Geographical distribution of the migratory caribou herds: the Rivière-George Herd, the Rivière-aux-Feuilles Herd, and the Monts Torngat Herd in Ungava Bay region (Cartography: Marc Girard, Département de géographie, Université de Montréal, 2013)

in several aspects of their way of life and their culture. As for several northern communities, subsistence hunting, fishing and trapping contribute greatly to their food and raw material needs. Among the numerous animals living on the tundra, the caribou has a very special status in the Naskapi culture and hunting this animal is of great importance to them. Since the eighteenth century, the Naskapi have been referred to in the literature as the "Indians of the Caribou" (Townsend 1911; Great Britain, Privy Council 1927; Speck 1935; Francis and Morantz 1983 cited in Mailhot 1986). They have developed an extremely rich and precise knowledge of this animal, including its feeding habits, its behavior under different circumstances (such as during the rutting season or when wounded), its physiological characteristics, and the diseases that affect it, its location and its migration routes. From talking to the Naskapi and through his own observations in 1893 and 1894, Low (1897) had identified three caribou herds that roamed northern Quebec. Caribou meat remains an important part of the Naskapis' diet (as for the Inuit, Cree and Innu peoples) because of its nutritional and energetic value, but also because of its cultural significance. The Naskapi use every part of the caribou for different purposes:

> When we shot down a caribou, we used all parts of his body. Even the bones were crushed and boiled into bone fat and broth that we drank. The marrow was eaten raw. We also used the powder of the burned bones to whiten tanned hides, which we rubbed with this powder. Then, we expanded the skins for the surplus powder to be blown by the wind. The skins of the caribou were used to make tents. We also made sinew with skin – thin strips that were used for fishing nets or fishing rods. These strips were also used to connect the different parts of sleds. The sinew was used to do many other things; for example, we tied our luggage when we wanted to travel. We also made rackets. We dried the meat and reduced it to very fine powder. We also used the shoes and made necklaces with the teeth and also games. When we made a drum, we used a lot of parts of the caribou.
>
> We also made toys for children with certain parts of the caribou. We respected this animal a lot because it allowed us to live and it was always present among us. (translation, Levesque et al. 2004–2008)

A place that is crucial to Naskapi culture is the Caribou Heaven. Legends that have been handed down in families for generations tell that it is a sacred place where the souls of dead caribous go. The Caribou Heaven, called atiuk weej in the Naskapi language, literally means "home of the caribou" and also plays a role in several other legends.

6 The Legend of the Caribou Heaven

According to Antoine Grégoire, the Montagnais guide for Jacques Rousseau, who went to the Québec-Labrador peninsula in the summer of 1951 to carry out botanical research, a Naskapi legend told of a place on the Koroc River called the 'Caribou Heaven' or the 'Door to Caribou Heaven' where the souls of dead caribou went. The soul of the animal was then given a new body, which allowed it to return to the forest.[11] Frank Speck, in 1935, also mentioned this myth:

> In the interior between Ungava Bay and Hudson's Bay is a distant country where no Indians will go under any consideration for the following reason. There is a range of big mountains pure white in color formed neither of snow, ice, nor white rock, but of caribou hair. They are shaped like a house and so they are known as Caribou House. One man of the Petisigabau band says there are two houses. In this enormous cavity live thousands upon thousands of caribou under the overlordship of a human being who is white and dressed in black. Some say there are several of them and they have beards. He is master of the caribou and will not permit anyone to come within some one hundred and fifty miles of his abode, the punishment being death.[12]

[11] Rousseau, J. (1953). Report on the Survey Carried Out in Northern Quebec Labrador. Montréal Botanical Garden: Montréal, 60.

[12] Speck, F. (1935), p.84. Op. cit.

The belief in the Caribou Heaven was initially spread throughout the Quebec-Labrador region and remains engraved in the memory of the Naskapi (and of the Inuits as well). According to Rousseau's theory, the legend may suggest that the ancestors of the Naskapi used to hunt on the Koroc River. He added however that « No Naskapi for generations has hunted in this place where now only occasional Eskimo venture in winter ».[13] No Naskapi alive today knows the exact location of the 'Home of the Caribou'. Only shamans have visited the Caribou Heaven, thanks to their preternatural powers of vision. Through the songs and legends that they created through their visions, they were able to order the caribou to come out of their house so that they could hunt them. Even if the Naskapi ancestors usually hunted near the Koroc River, the shamans forbade them to look for the mythic place out of fear that they would disturb the caribou.[14] Antoine Grégoire spoke of the Caribou Heaven as a place located "on the Koroc River, 75 miles upstream".[15] Another reference is given by Alain Hébert, who locates this site "[…] in the middle of the Koroc River valley, located on the west bank of this river, close to the junction with the André Grenier River".[16]

The legend of the Caribou Heaven or of the home of the caribou is very important for Naskapi culture. In the past, such beliefs provided ethical principles to guide the behaviour of Naskapi ancestors, whose survival depended largely on hunting caribou. For them, the responsible behavior that was encouraged by the legend and, in particular, the fact that all the parts of the slaughtered animal were used out of respect for the soul of the caribou, guaranteed that the caribou would come back to the hunter, thereby ensuring the survival of the Naskapi themselves. Today, several legends about the Caribou Heaven remain in the memory of the Naskapi Elders: some speak of people who got lost trying to find it; others give lessons to those who don't believe in it; still others speak of the belief in the soul of the mountain and in that of the caribou. Today the legend also serves the Naskapi as a tool to teach children the importance of treating not only the caribou, but all of nature, with respect. This myth helps the Naskapi (and the non-Naskapi) to understand that each person is a part of nature and has an important responsibility towards it.

[13] Rousseau, J. (1953), p. 56.

[14] Rousseau, J. (1953). Op.cit.

[15] Rousseau, J (1953), p. 60. Op.cit.

[16] Hébert, A. (2006) The Caribou Heaven in the Kuururjuaq Park: A legendary belief and maybe a sacred site. Ministry of Sustainable Development, Environment, and Parks: Québec. In Conseil de la nation Naskapi de Kawawachikamach (2007) (p. 4).

7 Protecting the Sacred Natural Site "Caribou Heaven" within the new Kuururjuaq National Park

Naskapi Elders held a meeting on February 14 and 15, 2007 in Kawawachikamach to determine if the Naskapi Nation would present a brief during the public hearing that would be held in Kangiqsualuujjuaq to recommend that the site that they refer to as "Caribou Heaven-Paradis des caribous" be designated as a sacred site. During group interviews conducted on this occasion, each Elder told the legend of the Caribou Heaven as it had been told to him by his father and grandfathers. A number of myths referring to the "home of the caribou" were told and recorded. The Elders agreed that it was important to protect the site of the Caribou Heaven because a substantial number of their myths refer to this site. Their fathers and grandfathers used songs and visions to call to the caribou whose souls were there. These accounts and stories convinced the Naskapi to request that the site of the Caribou Heaven be designated as a sacred site, and protected as such, so that the legends could be perpetuated and so that their children could learn them (Fig. 4).

On March 14 and 15, 2007, the Ministère du Développement Durable, de l'Environnement, et des Parcs du Québec, jointly with the Nunavik Environmental Quality Commission held public hearings in the village of Kangiqsualujjuaq concerning the project of the creation of the new Kuururjuaq National Park. Nearly 200 people attended these hearings, which were translated simultaneously into Inuktitut, French and English and broadcast throughout the villages of Nunavik via community

Fig. 4 Naskapi elders examine the documentation about the Caribou Heaven (© Photo: Blanka Füleki, 2007)

radio stations. An interim management plan, developed through a consultation process, presented the planned boundary lines, zoning proposals and a development concept.

On this occasion, the Naskapi Elders' advisory council and the Naskapi Nation Council recommended that the site of the Caribou Heaven, located within the boundaries of the proposed Kuururjuaq National Park, be designated as a sacred site and integrated into the future Park's educational facilities and that the Naskapi participate in managing the Park:

> We recommend that:
>
> - based on its cultural value for the Naskapi Nation, the Caribou Heaven be designated as a sacred area within the proposed Kuururjuaq Park;
> - that a Naskapi Elder at all times be a member of the committee responsible for managing the Park;
> - that an identification plaque in Naskapi, Inuktitut, English and French be installed at a convenient viewing point close to the agreed site of the Caribou Heaven;
> - that information about the Caribou Heaven be integrated into the cultural and educational facilities and materials of the future Park. To that effect, we undertake to share our knowledge. We shall commission a painting of the Caribou Heaven by a Naskapi artist as a gift to the interpretive centre.[17]

Following the public hearing, the site of the Caribou Heaven was declared an area of maximum protection because of its sacred character. The mere fact of designating a site as sacred does not ensure that it is protected; on the other hand, the fact that an area is designated as an area of "maximum conservation" does offer a guarantee of protection. In fact, the Park's zoning plan, which entered into force when the Park was created on May 21, 2009, establishes protection standards for the site.[18] No activity or sample-collecting are allowed in areas of maximum conservation. Scientific research and some educational activities can be authorized by the park director under certain conditions and upon presentation of a complete project description. However, these provisions do not apply to the Naskapi, the beneficiaries of the James Bay and Northern Quebec Agreement, who, because of the rights conferred on them, cannot have their access restricted.

Kuururjuaq National Park's permanent exposition, which is presently being developed, will address the legend.[19] The Naskapi have prepared a CD that contains several variations of the Caribou Heaven legend which they hope to see integrated into the park's educational program. Educational activities will make it possible to raise awareness among children and adults concerning the cultural and spiritual values related to the caribou, to Aboriginal knowledge of this animal and to the way that management practices that are part of Aboriginal myths and knowledge promote the preservation of the caribou and its habitat.

---

[17] Conseil de la nation Naskapi de Kawawachikamach (2007). Mémoire : Designation of the Caribou Heaven as a sacred area within the proposed Kuururjuaq National Park, 2 March, 4.

[18] Cossette, S. (2011). MDDEP (personal communication).

[19] Boulianne, M. (2011). Parcs Nunavik Parks. (personal communication).

8 Lessons Drawn

The protection of sacred natural sites of Indigenous and local communities is one of the oldest forms of culture-based conservation. These sacred natural sites often preserve fragile ecosystems and have long been an integral part of the identity and survival of Indigenous peoples and local communities, as well as the evolution of humanity as a whole. They are crucial places for the expression and transmission of culture, for biodiversity conservation and for the expression of spiritual values related to nature.

Cooperation agreements and land claims that guide the establishment and management of protected areas can offer the public organizations responsible for parks, Aboriginal communities, and other parties involved a way of working together to manage and protect the natural spaces essential to the culture of Aboriginals and to the continuance of their way of life, while achieving major conservation objectives at the same time. The case of the sacred natural site of the Caribou Heaven that we have dealt with in this text constitutes an innovative approach to conserving the biocultural heritage and to sharing the ecological, sociocultural, educational and economic wealth that protected areas bring.

Here are some of the lessons that can be learned from this case:

- The cultural and spiritual dimension of protected areas is crucial and must necessarily be taken into account;
- The protection of the sacred natural sites of Indigenous communities is one of the oldest forms of culture-based conservation. These sacred natural sites often preserve fragile ecosystems and have long been an integral part of the identity and survival of Indigenous peoples and local communities as well as the evolution of humanity as a whole;
- When a sacred natural site is located on land where there are plans to create a protected area, it must be recognized and managed accordingly by precisely recognizing the importance of cultural and spiritual values as an expression of the intrinsic relationship that indigenous people have with their land; just as it is necessary to recognize the Aboriginal beliefs, practices and knowledge through which the sites, cultures and resources associated with them have continued to exist;
- Allow community leaders to express their vision concerning the conservation and use of their ancestral lands;
- Ensure that neither time, patience nor confidence are lacking to develop a partnership among equals between the agency responsible for the administration of the Park, the communities and all the stakeholders.

References

Amend, T., Brown, J., Kothari, A., Phillips, A., & Stolton, S. (2008). *Protected landscapes and agrobiodiversity values* (Vol. 1). Heidelberg: IUCN and GTZ/Kasparek Verlag.

Argumedo, A. (2008). The Potato Park, Peru: Conserving agrobiodiversity in an Andean indigenous biocultural heritage area. In T. Amend, J. Brown, A. Kothari, A. Phillips, & S. Stolton (Eds.), *Protected landscapes and agrobiodiversity values* (Vol. 1, pp. 45–58). Heidelberg: IUCN and GTZ.

ARK. (2005). Projet de parc de la Kuururjuaq (Monts-Torngat-et-Rivière-Koroc). État des connaissances. Administration régionale Kativik, Service des ressources renouvelables, de l'environnement et de l'aménagement du territoire, Section des parcs, Kuujjuaq, Québec.

Berkes, F. (2008). *Sacred ecology. Traditional ecological knowledge and resource management.* Philadelphia: Taylor and Francis.

Bernbaum, E. (2002). Les montagnes sacrés. *Unasylva, 208*(53), 54–55.

Borrini-Feyerabend, G. et al. (2010). La diversité bio-culturelle conservée par les peuples autochtones et les communautés locales—exemples et analyses. Consortium APAC and CENESTA for GEF SGP, GTZ, IIED and l'UICN/CEESP, Téhéran.

Boudreau, S., Payette, S., Morneau, C., & Couturier, S. (2003). Recent decline of the George River caribou herd as revealed by tree-ring analysis. *Arctic, Antarctic and Alpine Research, 35*(2), 187–195.

Brown, J., Mitchell, N., & Beresford, M. (2005). *The protected landscape approach: Linking nature, culture and community.* Gland/Cambridge: IUCN.

Canadian Parks Council. (2008). Aboriginal peoples and Canada's Parks and Protected areas: Case studies = Peuples Autochtones, les Parcs et Aires Protegées du Canada: Etudes de Cas. Retrieved July 25, 2014, from http://www.parks-parcs.ca/english/cpc/aboriginal.php

CBC. (2014). *George River caribou population continues alarming decline.* 5 August 2014. URL: www.cbc.ca/news/canada/newfoundlandlabrador/george-river-caribou-population-continues-alarming-decline-1.2737260

Colding, J., & Folke, C. (2001). Social taboos: 'Invisible' systems of local resource management and biological conservation. *Ecological Applications, 11*(2), 584–600.

Courtois, R., Ouellet, J.-P., Gingras, A., Dussault, C., Breton, L., & Maltais, J. (2003). Historical changes and current distribution of caribou. Rangifer tarandus, in Quebec. *Canadian Field Naturalist, 117*, 399–414.

DELOS. (2014). Retrieved July 25, 2014, from http://www.med-ina.org/delos/

Descola, P., & Palsson, G. (1996). *Nature and society. Anthropological perspectives.* London/New York: Routledge.

Desponts, M. (2004). Les communautés végétales du territoire du projet de parc des Monts-Torngat-et-de-la-Rivière-Koroc, Nunavik, Québec. Rapport préparé pour l'Administration régionale Kativik. Ministère des Ressources naturelles, de la Faune et des Parcs, Direction de la recherche forestière.

Dignard, N. (2004) La flore vasculaire du territoire du projet de parc des Monts-Torngat-et-de-la--Rivière-Koroc, Nunavik, Québec. Report prepared for Kativik. Ministère des Ressources naturelles, de la Faune et des Parcs, Direction de la recherche forestière, Herbier du Québec.

Dudley, N., Higgins-Zogib, L., & Mansourian, S. (2005). *Beyond belief—Linking faith and protected areas to support biodiversity conservation.* Gland: WWF.

Ellen, R., & Fukui, K. (1996). *Redefining nature: Ecology, culture and domestication.* Oxford: Berg.

Francis, D., & Morantz, T. (1983). Partners in furs: A history of the fur trade in Eastern James Bay 1600–1870. McGill-Queen's University Press: Kingston et Montreal. In Mailhot, J. (1986). Beyond everyone's horizon stand the Naskapi. *Ethnohistory*, 33(4), 384–418.

Gillespie, D. A. (1983). The rock art sites of Kakadu National Park – Some preliminary research findings for their conservation and management. Australian National Parks and Wildlife Service, Special Publication, Canberra, 10.

Gomez, F., Hiemstra, W., & Verschuuren, B. (2010). A law on sacred sites in Guatemala. Policy matters, 17, Exploring the Right to Diversity in Conservation Law, Policy, and Practice, IUCN/CEESP, 116–120.

Great Britain, Privy Council. (1927). *In the matter of the boundary between the Dominion of Canada and the Colony of Newfoundland in the Labrador Peninsula, 1*. London: William Clowes and Sons.

Herrmann, T. M. (2006). Indigenous knowledge, values and management of the *Araucaria araucana* forest by the Mapuche Pewenche in the Chilean Andes: Implications for native forest conservation. *Biodiversity and Conservation, 15*(2), 647–662.

Higgins-Zogib, L. (2007). Sacred sites and protected areas: An interplay of place-views. In B. Haverkort & S. Rist (Eds.), *Endogenous development and bio-cultural diversity* (Compas series on worldviews and sciences, no 6, pp. 287–298). Leusden: IUCN.

Higgins-Zogib, L. (2008). The spiritual dimension of protected areas: Overlooked and undervalues. In Secretariat of the Convention on Biological Diversity (Ed.), *Protected areas in today's world: Their values and benefits for the welfare of the planet* (Technical series no. 36, pp. 50–57). Montreal: Secretariat of the Convention on Biological Diversity.

International Union for Conservation of Nature (IUCN). (2003). Vth world parks congress, recommendations. Retrieved July 25, 2014, from http://cmsdata.iucn.org/downloads/recommendationfr.pdf

Jean, D., & Lamontagne, G. (2004). Plan de gestion du caribou (Rangifer tarandus) dans la région Nord-du-Québec 2004–2010. Ministère des Ressources naturelles et de la Faune – Secteur Faune Québec, Direction de l'aménagement de la faune du Nord-du-Québec.

Koohafkan, P., & Boerma, D. (2006). Conservation and sustainable management of Globally Important Ingenious Agricultural Heritage (GIAHS). In Lee, C., & Schaaf, T. (Eds.), Proceedings of UNESCO-IUCN 'Conserving Cultural and Biological Diversity: The Role of Sacred Natural Sites and Cultural Landscapes' (30 May–2 June 2005) (pp. 80–88), Tokyo, Japan.

Kouami, K., Kossi, A., & Hamberger, K. (2005). Les forêts sacrées de l'aire Ouatchi au sud-est du Togo et les contraintes actuelles des modes de gestion locale des ressources forestières. *VertigO – la revue électronique en sciences de l'environnement*, 3(5). Retrieved July 25, 2014, from http://vertigo.revues.org/2456

Lévesque, C., Geoffroy, D., Polèse, G. with collaboration of the Elders of the Naskapis of Kawawachikamach. (2008). Les savoirs écologiques des Naskapis : caractérisation, utilisation, transmission – Projet de mobilisation des connaissances réalisé dans le cadre de l'initiative des Écosystèmes nordiques, Environnement Canada, 2004–2008. DIALOG, Institut National de Recherche Scientifique INRS, Montréal and Kawawachkamach, 5.

Low, A. P. (1897). Report on explorations in the Labrador peninsula along the Eastmain, Koksoak, Hamilton, Manicuagan and portions of other rivers in 1892–1895, *Annual Report of Geological Survey of Canada for 1895, 8*, 1–387.

Maffi, L. (2005). Linguistic, cultural, and biological diversity. *The Annual Review of Anthropology, 29*, 599–617.

Maffi, L., & Woodley, E. (2010). *Biocultural diversity conservation: A global sourcebook*. London: Earthscan.

Mallarach, J.-M. (Ed.). (2009). *Protected landscapes and cultural and spiritual values*. Gland: IUCN.

Mallarach, J.-M., & Papayannis, T. (Eds.). (2007). *Protected areas and spirituality*. Gland: International Union for Conservation of Nature (IUCN).

Nakashima, D., & Roué, M. (2002). Indigenous knowledge, peoples and sustainable practice. In Timmerman, P. (Ed.), Social and economic dimensions of global environmental change, Vol. 5 de *Encyclopedia of global environmental change* (Ted Munn [Ed in Chef], pp. 314–324). Chichester: Wiley.

Ormsby, A. A., & Bhagwat, S. A. (2010). Sacred forests of India: A strong tradition of community-based natural resource management. *Environmental Conservation, 37*(1), 1–7.

Ouellet, H. (1978). La vallée de la rivière Korok, une enclave forestière en toundra québécoise, Nord/North, 25, 6–11.

Oviedo, G., & Jeanrenaud, S. (2007). Protecting sacred natural sites of indigenous and traditional peoples. In J.-M. Mallarach & T. Papayannis (Eds.), *Protected areas and spirituality* (pp. 77–99). Gland: International Union for Conservation of Nature (IUCN).

Porter, I. (2011). Mystère dans la toundra. *Québec-Science*, March, 26–30.

Posey, D. A. (Ed.). (1999). *Cultural and spiritual values of biodiversity*. London: UNEP/ Intermediate Technology Publications.

Rousseau, J. (1953). *Report on the surveys carried out in Northern Québec-Labrador during 1951*. Montréal: Montreal Botanical Garden.

Secretariat of the Convention of Biological Diversity. (2004). Voluntary guidelines for the conduct of cultural, environmental and social impact assessments regarding developments proposed to take place on or which are likely to impact on, sacred sites and on lands and waters traditionally occupied or used by indigenous and local communities, Montréal. Retrieved, July 25, 2014, from www.cbd.int/doc/publications/akwe-brochure-fr.pdf

Sobrevila, C. (2008). *The role of indigenous peoples in biodiversity conservation: The natural but often forgotten partners*. Washington, DC: The World Bank.

Speck, F. (1935). *Naskapi: The savage hunters of the Labrador Peninsula*. Norman: University of Oklahoma Press.

Townsend, C. W. (Ed.). (1911). *Captain Cartwright and his Labrador journal*. Boston: Dana Estes and Company.

Turner, N., Ignace, M. B., & Ignace, R. (2000). Traditional ecological knowledge and wisdom of Aboriginal peoples in British Columbia. *Ecological Applications, 10*(5), 1275–1287.

UNESCO. (2003). Convention for the safeguarding of intangible cultural heritage. Retrieved, July 25, 2014, from http://www.unesco.org/culture/ich/index.php?lg=enandpg=00006

UNESCO, & UNEP. (2003). *Cultural diversity and biodiversity for sustainable development*. Nairobi: UNEP.

Verschuuren, B., Wild, R., McNeely, J., & Oviedo, G. (2010). *Sacred natural sites: Conserving nature and culture*. London: Earthscan.

Wild, R., & McLeod, C. (Eds.). (2008). *Sacred natural sites – Guidelines for protected area managers* (Best practice protected area guidelines series no. 16). Gland: IUCN/WCPA.

The Governance of Protected Areas in Greenland: The Resource National Park among Conservation and Exploitation

Daniela Tommasini

Abstract A significant part of Greenland has some forms of conservation status, and the National Park, created in 1974 with a surface of 972,000 km², mainly of inland ice and fjords, is the world's largest protected area. The National Park has a status of biosphere area under the Man and Biosphere Program (MAB). Strictly regulated for its access and allowed activities, e.g. recreational and outdoor activities are not authorized, and permission is needed, except for the population living adjacent to the Park, to be in the region but other activities, for instance mineral pits, are allowed.

Historically the establishment of protected areas has been based on either the protection of unique habitats or the concept of ecological representatively; nowadays climate change has become the primary challenge to the usefulness of protected areas as a conservation tool.

This chapter presents the case of Greenland, the environmental protection and the role of the local population in the governance of the national park seen from the local point of view as a resource for the socio-economic revitalisation of the adjacent community of Ittoqqortoormiit.

Keywords Protected areas • The National Park of Greenland • Governance and local population

Greenland has a total area of 2,175,600 km² stretching from Nunap Isua in the south (59.46° N) to Odaap Qeqertaa (83.40° N) in the north; nearly 90 % of the island's land mass is ice-covered and only around 410,449 km² is ice-free. Greenland straddles a boundary between two major climatic systems: north Atlantic maritime and polar arctic.

D. Tommasini (✉)
Faculty of Social Sciences, Multidimensional Tourism Institute (MTI), University of Lapland, Rovaniemi, Finland
e-mail: daniela.tommasini@ulapland.fi

T.M. Herrmann, T. Martin (eds.), *Indigenous Peoples' Governance of Land and Protected Territories in the Arctic*, DOI 10.1007/978-3-319-25035-9_7

Divided in four municipalities,[1] Greenland has 18 towns and 60 settlements, all located close to the coast. Its population is 56,370 (Statistics Greenland 2013), and the Inuit – the original population – is about 47,000 people. The capital is Nuuk which concentrate more than one third of the inhabitants (16,454. Statistics Greenland 2013). Roads do not connect towns and settlements, transportation of passengers and supplies of goods occur by sea or by air (helicopters and smaller airplanes). During wintertime dog sleds and snowmobiles are also used. The main part of the ice-free area of Greenland (99,8 %) is undisturbed by human activities, apart from the traditional hunting and fishing, these activities mainly do not influence the habitats but may have some impact on the main exploited species (Bugge Jensen and Christensen 2003). Fishing is the primary occupation in Greenland and is estimated to employ around 2500 individuals directly and another 3000 in the fishing industry (Statistics Greenland 2013). In addition to this, a number of people have occupations somehow connected with the fishery. Hunting has an influence on approximately 20 % of the population: it is the principal occupation in the more peripheral places in the West North and in the East of Greenland. Sheep, reindeers and some farming activities are to be found in the South of Greenland.

Greenland, which is part of the Kingdom of Denmark,[2] acquired Home Rule in 1979, and Self Rule in 2009 providing for full internal self-government (Act no. 473 2009); prior to this, the Greenlandic Council had legislative authority over certain hunting and fishing regulations and advisory capacity over other conservation issues (Fig. 1).

1 Environmental Protection and Protected Areas in Greenland

In 1962, the Greenlandic Council declared the sub-Arctic valleys of South-West Greenland with their unique and fragile "woods" as preserves (CAFF 1996: 24). In 1980 the Nature Conservation Act (Nature and Ancient Relics) enacted to safeguard and care for Greenland's natural scenic assets, so authorising protection of plant and animal species, as well as areas of land of high preservation or scientific value. Executive orders are used under the framework of this act to designate the protected areas: Northeast Greenland National Park (expanded by the Landsting Act No. 15, 1988) which is the world's largest National Park. The park includes two Ramsar

[1]On January 2009 the original 18 municipalities have been regrouped into 4. The Qaasuitsup Kommunia includes the original municipalities of Kangaatsiaq, Aasiaat, Qasigiannguit, Ilulissat, Qeqertarsuaq, Uummannaq, Upernavik, and Qaanaaq; the Qeqqata Kommunia includes the original municipalities of Maniitsoq and Sisimiut; the Kommune Kujalleq includes the original municipalities of Nanortalik, Narsaq, and Qaqortoq; Kommuneqarfik Sermersooq which includes the original municipalities of Ivittuut, Paamiut, Nuuk, Ittoqqortoormiit, and Ammassalik (Statistisk Årbog 2010. Nuuk: Grønland Statistik.).

[2]The Kingdom of Denmark is composed of three parts: Denmark, the Faroe Islands and Greenland.

Fig. 1 Greenland: towns, new municipalities and the National Park (Source: NunaGis, Ministry of Finance, Section of National Planning, Nuuk, Greenland)

sites[3] and was declared a Biosphere Reserve (MAB) in 1977. Animals and birds in the park are under total protection from outside visitors who require permission to visit it. Traditional harvesting activities are permitted by local communities (CAFF 1996: 24; National Park National Park 2010).

The Government of Greenland has the responsibility for environmental protection. The Ministry of Domestic Affairs, Nature, and Environment is responsible for environmental protection and nature conservation. The Environmental and Nature Protection Agency (APA) is taking care of: Environmental protection; Drinking water; Waste management; Particularly polluting enterprises; Soil contamination; Supervision of the environmental area; Coastal marine environmental protection; Wastewater; Supervision of the marine environment area. Nature protection; Nature conservation; Management of conservation areas; Supervision of the nature area;

[3] The Convention on Wetlands (Ramsar, Iran, 1971) is an intergovernmental treaty that embodies the commitments of its member countries to maintain the ecological character of their Wetlands of International Importance and to plan for the sustainable use of all of the wetlands in their territories (Ramsar Convention Secretariat 2010).

CITES[4] including issuing licences; The National Park of North and East Greenland. The management of land and marine mammals and birds is handled by the Ministry of Fisheries, Hunting, and Agriculture. The ministry is responsible for the management of international conventions on nature protection, specifically the Biodiversity Convention, the Ramsar Convention, and the Washington Convention.

In addition the ministry is responsible for the cooperation with Nordic and arctic parties and for the management of protected areas, including the World Heritage area in Ilulissat (Ministry of Housing, Nature and Environment. www.nanoq.gl).

Greenland, as part of the Danish commonwealth, has joined the Washington Convention, which covers trade in animals and plants threatened with extinction (www.cites.org); The Ramsar Convention protects areas for birds (www.ramsar.org); The Biodiversity Convention concerns the protection of the total biological diversity (www.cbd.int). In the Nordic context, Greenland participates in nature protection, the efforts being concentrated in the working group on terrestrial ecosystems under the Nordic Council of Ministers (www.norden.org; www.nanoq.gl); and in the Aquatic Ecosystems Working Group and the Waste Working Group under the Nordic Council of Ministers (www.norden.org) (Ministry of Housing, Nature and Environment).

The Kingdom of Denmark Strategy for the Arctic 2011–2020, focus on the Kingdom's strategic priorities for future development in the Arctic towards 2020. The aim is to strengthen the Kingdom's status as global player in the Arctic (Strategy for the Arctic 2011: 11) and its nature and environment must be managed based on the best possible scientific knowledge and standards for protection, and international cooperation in this endeavour must be promoted (Strategy for the Arctic 2011: 43).

A significant part of Greenland has some form of protection, there are 964,795.4 km^2 of protected land of which 18 % (176,917.4 km^2) is ice-free land, 12 % (115,793 km^2) is sea, and 70 % (672,085 km^2) is ice. The world's largest National Park is located in sparsely populated Northeast Greenland and by large part it consists of Greenland Ice Cap; the park encompasses about 32 % of Greenland's ice covered area: the 43 % of its ice-free area incorporates two Ramsar sites (Salathe 2009; Bugge Jensen and Christensen 2003). In addition to the National Park, there are six other protected areas in Greenland (in accordance with Home Rule legislation no. 11 of 12 November 1989), covering about 8.100 km^2.

Beside the National Park, the following areas have in Greenland a conservation status:

- Qinnguadalen by Nanortalik has the largest birch forest of Greenland
- The island of Uunarteq by Nanortalik for the hot springs.
- The Ikka Fjord by Ivittuut for the unique limestone pillars of the fjord
- The island of Akilia at Nuuk for the old geological deposits

[4] CITES, the Convention on International Trade in Endangered Species of Wild Fauna and Flora is an international agreement between governments. Its aim is to ensure that international trade in specimens of wild animals and plants does not threaten their survival (www.cites.org).

- Arnangarnup Qoorua at Maniitsoq for the unique flora and fauna of the area, and cultural interests
- The heather field at Qeqertarsuaq for the unique flora in the area surrounding the Arctic Station
- Melville Bay between Upernavik and Qaanaaq as an important breeding area for narwhals
- Ilulissat and the Ice fjord in the North-West of Greenland. Protected area from 2003 and Unesco World Heritage Site from 2004
- Kitsissunnguit (Grønne Ejland) in the Disko Bay for the abundant bird life
- Austmannadalen at Nuuk for the landscape and for the stock of wild sheep

The nature reserve in Melville Bay was established primarily to protect wildlife associated with the marine environment, narwhals, beluga whales and polar bears. In one part of the reserve, professional hunters from counties adjacent to the reserve are allowed to conduct traditional hunting trips. However, travel in the remaining part of reserve is not allowed. Other protected areas have been established for scientific research, such as Lyngmark on Qeqertarsuaq, wherein travel and hunting are not prohibited. In the valley of Arnangarnuup Qoorua animals and plants are protected year round and only travel by foot is permitted. Qinngua Valley, the largest area in Greenland with birch forest, is protected to preserve this unique environment; the entry to the area is allowed but hunting and other activities that can harm the environment are not. The small island of Akilia is protected to preserve its geological formations. The Government can give permission for collecting rocks and minerals for research purposes (Bugge Jensen and Christensen 2003). In order to preserve the unique Ikka Columns that rise off the sea floor, the Ikka Fjord is protected since year 2000. There are restrictions on sailing with motorized boats, fishing with trawls, or other implements that may damage the columns, and collection and destruction of columns.

In addition to these protected areas, there are 11 areas in Greenland designated as Ramsar sites. Two of these are in the National Park. The remaining nine cover a total area of 12,500 km^2. The aim of the Ramsar Convention is to protect important wetland and coastal ecosystems (Boertmann et al. 2009). With the accession of the convention, the participating countries commit to protecting the ecological value associated with designated sites and ensure their sustainable use. Outside the National Park (which includes two Ramsar sites), the nine Ramsar sites make up 61 % (1,250,000 ha) of the surface of Greenland's protected areas (2,060,000 ha, or 5 % of Greenland's ice-free area outside of the National Park). This means that the category "Ramsar sites" covers a substantial part of Greenland's protected areas (Salathe 2009). Besides, Greenland has appointed 11 Ramsar areas with a total of 15,457.5 km^2, however no legislation has been established to protect these areas, and hunting, fishing and access is regulated through the same rules as outside the Ramsar areas (Due and Ingerslev 2000).

2 The National Park of Greenland

The National Park of Greenland, an emblematic representation of high arctic nature, was created in 1974 and extended towards west in 1988, covers now a surface of 972,000 km², mainly of inland ice and fjords. It is extended from the northern part of the inhabited places of Ittoqqortoormiit in East Greenland to the north-eastern part of Qaanaaq in North of Greenland (Statistics Greenland – Nature 2013) (Fig. 2).

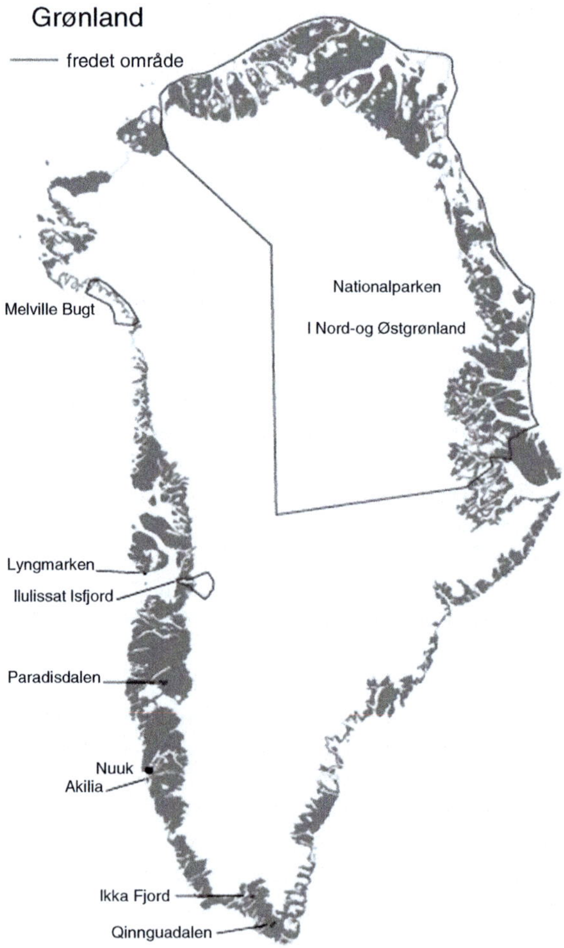

Fig. 2 The National Park and other protected areas (Source: Ministry of Domestic Affairs, Nature and Environment, Nuuk, Greenland)

According to statistics, there is officially no permanent human population; however, some scientists and military are stationed there.[5] Near the coastal zones it is estimated that a population of 5,000–15,000 musk ox as well as polar bears and walrus can be found, together with other mammals such as arctic fox and arctic hare. Marine mammals include ringed seal, bearded seal, harp seal and hooded seal as well as narwhal and white whale. Species of birds which breed in the park include great northern diver, barnacle goose, pink-footed goose, common eider, king eider, gyrfalcon, snowy owl, ptarmigan and raven (Statistisk Årbog 2013).

The National Park also has a status of biosphere area under the Man and Biosphere Program (MAB) of UNESCO. According to the regulation, recreational and outdoor activities are not authorized in the park. Only scientific expeditions after receiving special permission from the authorities are allowed. Other activities, for instance mineral pits, are allowed. The protected area hosts also the Sirius Patrol, a military detachment with duties of territorial control. Its conservation status is as a National Park but it does not comply with IUCN definition. It is a Biosphere reserve under UNESCO Man and Biosphere Programme. Internationally recognised as a National Park, however since the permission to carry out mineral exploitation in the National Park (1994) is allowed, Greenland no longer complies with criteria for the IUCN protection category II, National Park.[6]

In the National Park, everyone, except for individuals from Ittoqqortoormiit and Qaanaaq (Avanersuaq) regions, needs a permit to be in the region. Hunters from the two regions are allowed to hunt in the National Park. The Sirius Patrol and personnel from the permanent military and weather stations are permitted to hunt seals for feeding the dog teams and to fish for private use.

According to the Home Rule Authority (Ex. Ord. 7/1992 and 5/1999) the purpose of the National Park is to conserve the wilderness of the region and at the same time allow research and public admission. Protection of landscapes, flora, wildlife, prehistoric remains and other cultural relics of the past is the overall objective.[7]

As in Chapter I (General regulations, § 1) The National Park is administered by the Greenland Government *Landsstyret* (§ 24). All hunting is forbidden (Chapter II,

---

[5] Daneborg (12 people) is the headquarters of the Sirius Patrol; Danmarkshavn (8) is a civilian weather station; Station Nord (5) is a military base; Mestersvig (2) is a military outpost with 1,800 m gravel runway; Zackenberg (0) is a summer-only research station can host on average 20 scientists and station personnel; Summit Camp (4) is a research station on the Greenland Ice Sheet (Cf. Statistisk Årbog 2009).

[6] IUCN has defined a series of six protected area management categories, based on primary management objective. Category II National Park: protected area managed mainly for ecosystem protection and recreation. Definition Natural area of land and/or sea, designated to (a) protect the ecological integrity of one or more ecosystems for present and future generations, (b) exclude exploitation or occupation inimical to the purposes of designation of the area and (c) provide a foundation for spiritual, scientific, educational, recreational and visitor opportunities, all of which must be environmentally and culturally compatible.

[7] Executive Order no. 7 of 17 June 1992 from the Greenland Home Rule Authority concerning the National Park in North and East Greenland, as amended by Executive Order no. 16 of 5 October 1999. In accordance with § 16, Sec. 2 in: Act of Landsting no. 11 of November 12th 1980 on Nature Preservation.

Fishing and hunting, § 2) however personnel of the permanent stations[8] can fish and shoot seals for their own consumption (§4). Persons permitted to enter the National Park (Chapter VII, Admission and control arrangements, § 21) are: persons with permanent residence in Avanersuaq (Thule) and Ittoqqortoormiit (Scoresbysund) municipalities and with close connection to the Greenlandic society; persons performing duties for public authorities; and persons who work at the stations. These persons are authorised to conduct traditional hunting in the National Park, under compliance with wildlife conservation regulations for the municipality of Ittoqqortoormiit, and pending that they possess a valid certificate stating that hunting is their primary livelihood (§ 22).

A permit to preliminary investigation, exploration, and exploitation of mineral resources in the National Park can be granted in accordance with the stipulations mentioned in the "Law of Mineral Resources in Greenland" (§ 25), pending comments from the "Directorate for the Environment and Nature" concerning special stipulations relating to the individual permit. For the "Geological Survey of Denmark and Greenland" and any other publicly authorized institution operating within the National Park, stipulations are issued in a similar way (see Chapter VII, Admission and control arrangements, § 25). In the area of Maniitsoq (West Greenland) Arnangarnup Qoorua in Angujaartorfiup Nunaa is declared as a protected area; "... is designated as a protected area, owing to its beautiful landscape and importance both culturally and scientifically" (§ 4 and § 9 of Act of Landsting no. 11 of November 12th 1980). As in § 6: "Notwithstanding the conditions laid down in this Act, permission to carry out pilot studies, investigations and exploitation of the mineral raw materials in the protected areas, may be granted, according to Act no. 844 of December 21st 1988 concerning natural raw materials in Greenland".

The "Mineral Resources Act" of 2009[9] contains a number of parts and rules concerning the "protection of the environment, the climate and nature as well as on prevention, limitation and combating of pollution and other negative impact on environment, the climate and nature".[10] The Greenland Government "must attach importance to the consideration for avoiding impairment of nature and the habitats of species in designated national and international nature conservation areas and disturbance of the species for which the areas have been designated when making a decision on the granting of a licence" (...) "It is decided under national and international law applicable in Greenland which areas are national and international nature

[8]The following are included under permanent stations: Mesters Vig Airport, Ella Ø, Daneborg, Danmarkshavn, Station Nord, Kap Moltke, Brønlundhus (Chapter VIII, § 26). A permanent station includes all buildings and facilities necessary for its operation, incl. lakes and rivers for drinking water supply, harbour, approach zones and air strips on land, lake ice and sea ice, dumps, and the commonly used roads or tracks to the above mentioned localities (Chapter VIII, § 26 Sec. 2).

[9]Greenland Parliament Act no. 7 of December 7, 2009 on Mineral Resources and Mineral Resources Activities.

[10]The Mineral Resources Act, 7/2009. General Explanatory Notes, 5: Consequences for the environment and nature: 29.

conservation areas and which national and international rules apply to such areas".[11] A licence may be granted only if "the activity or the facility does not damage the integrity of a national or international nature conservation area" (...) nevertheless "if an impact assessment shows that the project does not damage international nature conservation areas, there is nothing to prevent the granting of a licence for the project".[12]

3 Meanings and Values of a Protected Area

Protected areas are a cultural construction, and have a relatively long history in Europe, it was during The Renaissance that protected areas were as first considered, when the hunting grounds were set aside for nobles and aristocratic. Later protective areas acquired a more democratic sense, were open for the public and considered a collective good for the benefit of the community, already providing first forms of tourism. The 60ies saw a growing concern for protecting nature, and following these ideas also in Denmark the growing commitment for protecting nature gave way to the idea and realisation of the National Park of Greenland, the biggest protected area in the word.[13] From the 1960s onwards the all idea of nature was revisited, ecology became popular, and if at the beginning the meaning of protecting an area was more "to set aside", later on it became more sophisticated including in the debate issues like the cultural values and the protection of landscapes.

Usually the decision to protect an area is made from the government, regulated by law (almost every country has a protected areas legislation) and with clear rules. The area under protection has normally an interest from the point of view of the environment, the landscape, also cultural landscape, and is open to the public (this latter is not the case for the National Park of Greenland).

For long time a protected area has been seen as something "to protect" from external influences. In recent years, especially in European regions, the concept of protecting and area has changed, and areas object of preservation can have also human activities, even with seasonal settlements. Recreational activities are now one of the topics of protected areas, which are becoming more and more popular because of the recreational and educational offer. Arctic protected areas, like protected areas throughout the world, hold value for society. Just as there is a wide diversity in protected areas, there is likewise a wide diversity in the values they protect and represent (Pagnan et al 2004: 1).

[11] The Mineral Resources Act, 7/2009, Explanatory Notes to the individual provisions of the Bill, section 60: 92.

[12] The Mineral Resources Act, 7/2009, Explanatory Notes to the individual provisions of the Bill, section 61: 93.

[13] The first National Park in Denmark was created in 2008, in the north of the Jutland peninsula. Retrived from http://www.danmarksnationalparker.dk

Not always local population welcome the creation the establishment of a protected area, considering protection merely as a limitation for the access, and hindrance to traditional activities on that area. Local populations rarely can see the purpose of a protected area, perceived as a hindrance to many activities and they do generally not completely value the institutional meanings of protecting an area adjacent to the community. Generally, the local population's idea of resource, is of something to benefit of, to take advantages of, to profit of, for instance with some seasonal permission for hunting in limited areas within the national park, or with some locally based tourism business providing tourists' excursions in the protected area and locally based companies for the logistic support of scientific expeditions.

The decision of establishing a protected area has economic implications as well; around the world, protected areas help generate money and jobs. Even if protected areas do not always offer the same sort of short-term monetary benefits as, for example, the oil and gas industry, however, the benefits offered by protected areas are, instead, long-term, more sustainable and may be better suited to the aspirations of many northern residents (Pagnan et al. 2004: 13). They attract tourists as well as scientists, and the business sector. They can provide employment for local communities so after initial suspicion, local population may recognize the economic benefits of the protected area, for instance from a touristic point of view where communities can profit for the increasing of services and infrastructures in the area and for the creation of new jobs.

There are no universally agreed definitions of protected areas, but two which are widely used are that for protected areas in general in the UN Convention on Biological Diversity (CBD) and the IUCN definition for marine protected areas (MPA). According to the most recent update of the protected areas in the Arctic, made under CPAN in 2004, almost 20 % of the Arctic land mass is judged to have protected area status in terms of IUCN categories. This is greater than the global average, which stands at some 11.5 %. The same does not apply to marine areas, since little of the Arctic marine environment has been designated as marine protected areas (MPA). The 2004 CPAN update argues that the Arctic is not alone here; according to statistics compiled for the 2003 World Parks Congress, the rest of the world faces the same challenge, with less than 2 % of the marine and coastal environment on the globe managed as protected areas or conservation zones (Koivurova 2009: 45).

Historically the establishment of protected areas has been based on either the protection of unique habitats or the concept of ecological representatively; nowadays climate change has become the primary challenge to the usefulness of protected areas as a conservation tool. These complex ecosystems are vulnerable to climate-driven ecological change, industrial development, and resource exploitation (Barry and McLennan 2010). The first protected areas in the Arcticwere established in Sweden and Alaska at the beginning of the twentieth century. The area under protection remained low until the 1970s when it began to increase significantly with additions of large areas such as the Greenland National Park. By 1980, 5.6 % of the Arctic was classified under some degree of protection. This has steadily increased until today where 11 % of the Arctic 3, about 3.5 million km^2, has protected status in 1127 protected areas. Of course, the nature of protection and governance of these areas varies throughout the circumpolar region, and there are varying levels of protection within countries (Barry and McLennan 2010: 97).

The Ministry of Domestic Affairs, Nature, and Environment of Greenland participate in the international environment protection work through working groups such as the Arctic Monitoring Assessment Programme (AMAP) and the Protection of the Arctic Marine Environment (PAME). The working groups monitor transboundary pollution and climate change and work to reduce pollution in the arctic countries. Greenland is member of international conventions and agreements, either through Danish ratification or by becoming a signatory itself, such as the Convention on Biological Diversity (CBD) ensuring protection of biodiversity through sustainable use and monitoring, local involvement and other issues; the Convention on International Trade in Endangered Species of Wild Fauna and Flora (CITES) regulating and monitoring trade of wild species, their parts and products; the International Whaling Commission (IWC) which make decisions on whaling quotas (for aboriginal subsistence whaling) and guidelines for best practices for whaling and for the protection of whales. Moreover the Ramsar Convention on Wetlands protecting internationally important wetlands with unique aggregations of birds and other wildlife; the Agreement on conservation of polar bears protecting polar bears in the circumpolar countries; Arctic Council which with the Conservation of Arctic Flora and Fauna (CAFF) set common goals for management of arctic flora and fauna; the North Atlantic Marine Mammal Commission (NAMMCO) issuing specific management recommendations in terms of hunting levels and protection. Furthermore the Canada/Greenland Joint Commission on Conservation and Management of Narwhal and Beluga (JCNB) issuing specific management recommendations in terms of hunting level and protection and the World Conservation Union (IUCN) giving management advice and cooperating on information exchange. The UNESCO World Heritage Convention ensures classification and protection of unique nature and cultural sites. Greenland is also a member of the Nordic Council and the Nordic Council of Ministers where cooperation in terms of environmental protection and nature protection also takes place.

The Greenland Ramsar sites were included in the Ramsar[14] list of international important wetlands in 1988. In 1997 the Greenland government decided to apply for Ramsar status for eleven sites in Greenland. In January 1998 these sites were acknowledged by the Ramsar secretariat and included in the Ramsar list of international important wetlands. More than 13,400 km^2 of Arctic ecosystems are located both in East Greenland (three sites) and in West Greenland (eight sites) and include wetlands in marine, tidal and fresh-water environments. Compared with the other Ramsar sites in the world, the Greenland sites include a higher proportion of marine habitats (Frazier 1999), and none of the sites comprise permanent human settlements.

[14] The Convention on Wetlands (signed in Ramsar, Iran, 1971) is an intergovernmental treaty which provides the framework for national action and international co-operation for the conservation and wise use of the Worlds wetlands and their resources (Cf. Egevang and Boertmann 2001).

4 Governance and Local Population: The Resource National Park as Opportunity for the Socio-economic Revitalisation of the Adjacent Town of Ittoqqortoormiit

The region of Ittoqqortoormiit, in the East Coast of Greenland, is located at approximately 70°31′N, 22°00′W near the mouth of the Kangertittivaq / Scoresby Sound, and covers an area of 235,000 km^2 along the Denmark Strait and the Greenland Sea. North, it borders on the Greenland National Park, the largest in the world, and to the south west it borders with the Ammassalik region.[15] Although Ittoqqortoormiit is situated only half way up the Coast of East Greenland, the climate is more typical of the high-Arctic. A cold south flowing current, emerging from the Arctic Ocean sweeps by the northeast Greenland coast and helps to control regional weather patterns (Geografi, Klima og Natur 2013).

Located at the arm of the world's largest fjord system, Ittoqqortoormiit has a very remote geographical situation, one of the most remotes in Greenland. Ittoqqortoormiit has 452 inhabitants; there are two settlements, Ittorisseq/Cape Hope and Uunarteq/Cape Tobin, uninhabited since 2006.[16] (Cf. Statistics Greenland). All three settlements are situated at the southern tip of Liverpool Land along the northern entrance to Scores by Sund (Figs. 3 and 4).

The name Ittoqqortoormiit means "those who live in the large houses". Established in 1924 after the Ejnar Mikkelsen expedition with the purpose of moving part of the inhabitants from Ammassalik (Tasiilaq), which lies 900 km south, had reached a population limit to what the nearby hunting areas could supply. Another reason was geopolitical: after some territorial disputes with Norway, the kingdom of Denmark, which Greenland was a colony at that time, decided to set an outpost far north.[17] Originally the town was flanked by some the settlements, which

[15]With the establishing of the new municipalities (Jan 1st, 2009) all these borders have a pure geographical sense, Ittoqqortoormiit and Ammassalik are now part of the municipality of Sermersooq, which comprises also the main town, Nuuk, Paamiut and Ivittuut.

[16]As for the population dynamics in 1990 the total number of the settlement's inhabitants was 84; in 1994 the number decreased at 40 units and ten years later, in 2004 there were only 9 inhabitants. At the same time, Ittoqqortoormiit in 1990 had 554 inhabitants; in 1994 the number decreased at 524 units, and in 2003 it decreased at 519 and today is equal to 452 (Cf. Statistics Greenland 2013 and other various years).

[17]About the year 1000 A.D. Vikings came to Greenland, and established two settlements in West Greenland, and are known to have used high mountains in East Greenland as landmarks. There is circumstantial evidence of direct contact between the Vikings and the Greenland population of the Scoresby Sound region, in the form of silver buttons and beads found in Inuit graves. The area, as testified by ruins and other archaeological remains, had been home to a dense population of Inuit in the past. William Scoresby senior and his son (also William Scoresby) sailing in East Greenland waters, reported observations of land between 70° and 74°N in their whaling logs in 1817 and 1821. In 1822 Scoresby senior (on the Fame) and Scoresby junior (on the Baffin), together with 20–30 other British whalers, were on numerous occasions close to land. Scoresby junior named Scoresby Sund after his father. Harald Olrik already in 1911 proposed the foundation of a settlement in the unpopulated tracts of Scoresby Sund. The project was brought to realization in 1924 due to the interest and influence of Ejnar Mikkelsen. About 85 Greenlanders arrived in 1925,

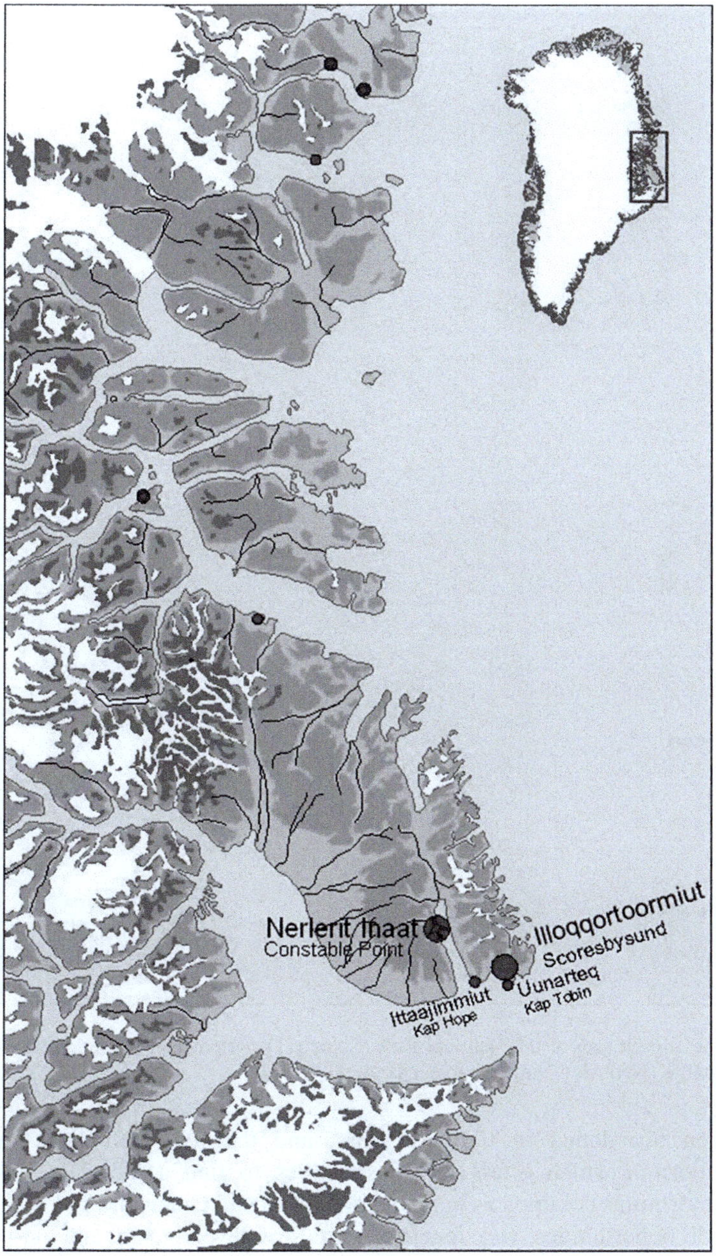

Fig. 3 The Ittoqqortoormiit area (source: Greenland Tourism, Nuuk, Greenland)

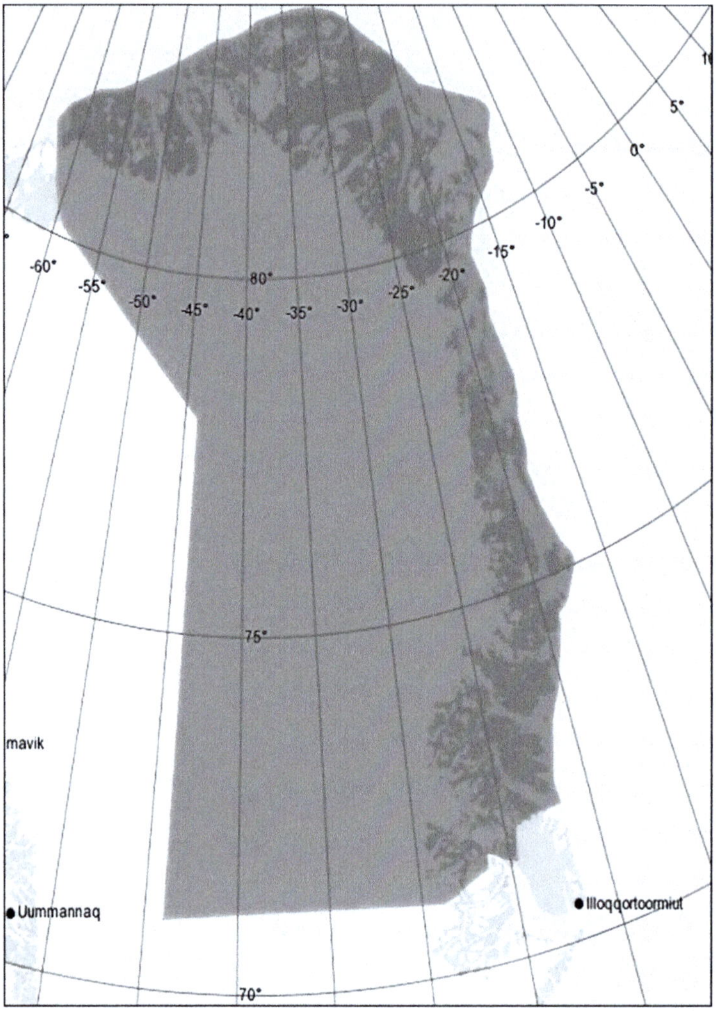

Fig. 4 The current size of the National Park (Source: Departementet for Indenrigsanliggender, Natur og Miljø (NNPAN), Nuuk, Greenland)

have been abandoned in the last years, and the few inhabitants moved to Ittoqqortoormiit which offers few possibilities to find a job. The situation in Ittoqqortoormiit is common as in many other places in Greenland and in the Arctic, lack of job opportunities, very few possibilities of development, and out-migration

which is the foundation of Ittoqqortoormiit / Scoresbysund. Houses were built at Kap Stewart, Kap Hope and Kap Tobin for the Greenlander hunters and their families. The settlement was encouraged by the colonial power that at the time had a growing interest in Northeast Greenland, and at the same time, the colonization was intended to improve declining living conditions in Tasiilaq, from where the settlers came mainly from. The settlers (Greenlanders) soon prospered on the good hunting conditions of the new area, which was rich in seals, walruses, narwhals, polar bears and arctic foxes (Tommasini 2011).

is quite significant, an average of 100 people out of 500 left in the last four years (Statistics Greenland 2009) and not only for educational purposes.

Apart from the service jobs (shop, school, the little hospital, kindergarten and the elderly people home), very few are the chances to get a permanent job. The local fish factory has almost no activity, except for the short time where hunters can sell the mattaq (whale skin). Except the small amount of remunerated jobs the local economy is a mix of seasonal tourism activities, subsistence hunting and social assistance. Hunting activities are the key part of the local (subsistence) economy, and for many the main and only activity remains subsistence hunting. Local hunters have for generations lived from whale and polar bear hunting, still a significant cultural-economical factor in the area. Flesh and by-products play a direct part in the economy of the hunting families. Income is gained by trading these products, but these options are seasonal and variable. Ittoqqortoormiit lies near large populations of shrimp and Greenland halibut, but the presence of sea ice prevents exploiting these resources year-round, and as a result fishing has never been extensively developed in the municipality (Tommasini 2009).

Tourism, on the other hand, is growing in importance. At present Ittoqqortoormiit, a rather small town has limited infrastructure and limited capacity. The area is still remote; the closest neighbouring town in Greenland is Tasiilaq about 800 km to the southwest. Ittoqqortoormiit has no direct connection with the rest of Greenland, it is only reachable by plane with a direct flight from Iceland to Constable Pynt airport, and then with helicopter to Ittoqqortoormiit, (Scoresbysund) situated on the coast of East Greenland. There are usually two weekly flights from the end of March to the end of October, and one weekly connection during the rest of the year. Several expedition cruise ships also call at Ittoqqortoormiit as a result of the growing cruise ship tourism in all Greenland in the last years (Tommasini 2009).

Tourism appeal due to the remoteness, impressive landscape and the activities related to tourism seems to be an important component in the economy of the place. Tourism can generate good money for the hunters offering dog sledge tours to the tourists. In town there are about 50 dog sledges and rough calculations indicate that in 2005 an average one million Danish crowns, were earned during the whole tourism season, that means dog sledding and sailing (informant 2, 2009).

The place offers several activities all year round, for instance dog sledding, sailing, and hiking. The peak season is between March and May for dog sledding and between August and September for sailing, October is the more favourite for cruise ship tourists. The number of visitors is around 150 for dog sledding, 180 for sailing and more than 1000 for cruise ships (Tommasini 2011).

The tourism offer in Ittoqqortoormiit is primarily represented by an impressive landscape, a rich cultural heritage still traditional, and a national park which represents a remarkable appeal from the point of view of tourism, and protected areas are, from a point of view of tourism, appealing. It is commonly said that tourism needs protected areas and protected areas need tourism. In the case of Greenland and its National park we cannot actually say completely so. The image of a protected area such a park can be conveyed in different ways; in this case the message seems to be this of the biggest national park in the world, where nature and landscapes will be kept back from almost any influence. Thus, an image can change, for instance the population living at the border with the national park declared (Tommasini 2011) that, having already some little tourism activities going on, they would like to "use" the resource as opportunity to stimulate and reinforce the local economy. Perhaps, when the area to protect, in a region almost filled with ice, was established, among the purposes there was not the idea of using the area also for educational and recreational use.[18] Things can always evolve, and it may be possible to convert (at least) the image of a protected area from a static and almost untouchable place to something more active. A renewed and fresher image of the park can contribute to the revitalisation of the nearby region, offering new opportunities for recreation in the area adjacent to the National Park. The Ittoqqortoormiit region is blessed by stunning sceneries, by the biggest fjord system in the world and Jameson Land bordering the Park, hence famous for its beautiful landscape. Promoting educational and leisure activities can give way to significant outcomes for the local community (Tommasini 2011).

In Ittoqqortoormiit the local population's idea of resource is of something to benefit from and to take advantage of, like the allowed seasonal permissions for hunting (according to quota) in limited areas within the national park. Some locally based tourism business providing tourists' excursions in the protected area, and locally based companies for the logistic support of scientific expeditions may significantly contribute to boost the local economy. The challenge is to develop a tourist product in fact already existent, the park. This needs investments, most of all, in human capital. Trained local guides are winning cards for the future of the community, which can thus avoid having guides coming from outside. To become a guide is a task that may require an important investment in time and money (Nickels et al. 1991), and at the early stages of development it may be too costly for a small community. However a guide can be everyone in the community that knows the place and the surroundings, and to some extent a foreign language. This is a crucial factor, and presently few of the hunters have some knowledge of the English language in Ittoqqortoormiit.

In Ittoqqortoormiit the role of the local population in the decisional process still remains quite marginal as observed during field work in 2009 (Tommasini 2011) and, as reported by an informant, "in August 2010 during a public meeting about the park in Ittoqqortoormiit, the Man and Biosphere (MAB) was explained to the audience, consisting of 30–40 people. We (informant) explained what it is the park and why is good to have it. People did not know about and was sceptic, then eventually agreed to do something to "sell" Ittoqqortoormiit to tourism. The town is dying now. The government says we can have 6 to 8000 tourists a year if an airport will be built closer to Ittoqqortoormiit and we can combine the resource National Park with the town. Mining is in standby because of the economic crises. We have a very big problem because of no hunting activities; we have to find another way. There are no seals and right now (November) it should be a lot. No snow, no possibilities to go by sledge and reach the musk-ox hunting fields. Tourism went dramatically down in

[18]Even some activities are allowed. In the specific case the authorities consent some mineral exploitation. For the pit activities even a land strip, now almost abandoned, was constructed in the zone that fall inside the National Park area.

the last years; there are no infrastructures, and no restaurant. If the airport will be build close to the town then a hotel with 20 rooms and a restaurant will be built. Everything depends on the airport. Lack of jobs in Ittoqqortoormiit, the population dropdown is nearly of 100 units, and this is a lot for a small community" (informant 1, 24.11.2010).

In general in Greenland there is broad political consensus to develop the mineral and hydro-carbon sector into one of the mainstays of the economy (Convention of Biological Diversity of Greenland 2004) and Kommuneqarfik Sermersooq, where Ittoqqortoormiit is part of, is also in favour of expanding the Man and Biosphere area including Ittoqqortoormiit and the surrounding fjord complex in order to fulfil the MAB requirements. The plan is to use the town centre and base for the logistics of travel to and from the park. Thoughts and wishes have been put forward of a small landing strip near by Ittoqqortoormiit in order to serve and increase the tourist industry (Olsen 2010).

5 Conclusions

Greenland's approach to nature conservation is heavily focused on species conservation and sustainable use of living resources. It uses a combination of a traditional area protection approach and a huntingfishing and harvesting regulations, and the weight placed on user group restrictions for species conservation is higher than in most other national systems (CAFF 1996: 24). Generally, the ecological information from the Greenland sites is very limited, and dates in some cases back to when or even before the sites were designated. Surveys and studies have been scattered and mainly of opportunistic character, and no monitoring programmes have been established. Through the 1990s information available on the Greenland Ramsar sites were collected, and subsequently published in 1990, 1993 and 1996 together with similar information from the Danish Ramsar sites by the National Forest and Nature Agency in Denmark (Egevang and Boertmann 2001; Jepsen et al. 1993).

Currently, government-monitoring efforts are focused on harvested resources,[19] threat monitoring in response to pressures from industrial development, including mining, oil and gas exploration, and increased shipping (Livingston 2011). However,

---

[19] Greenland's monitoring programs currently include the following:

- Greenland Ecosystem Monitoring (GEM) at two sites, of which one is at the Zackenberg Research Station in Northeast Greenland National Park and the other near Nuuk (not within a protected area); Greenland Institute of Natural Resources, monitoring of harvested species (some in protected areas), threat monitoring (some in protected areas), and local monitoring by non-scientists (some in protected areas). The Zackenberg monitoring program has been underway since 1995 and includes monitoring on five themes: climate, marine, geological, glacial, and biological. The latter includes monitoring of the dynamics of a large variety of organisms and biological processes in the local ecosystems;
- Monitoring of harvested species, in some cases dating back over 100 years, including narwhal, other whales, walrus, harbour seal, polar bear, musk ox, reindeer, fox, hare, guillemot, eider,

Ittoqqortoormiit, the nearest settlement to the national park, has so far a marginal role in the decision process. The roles of indigenous peoples in national designation processes varies widely, ranging from co-management of protected areas in Canada, to a special emphasis on the protection of indigenous cultures and ways of life along with species and habitat as is found in Russia and Greenland (Cf. Belikov and Legare 1996: 9). The protection of areas in the Arctic will have the largest impact on the Arctic indigenous peoples and on local communities nearby or adjacent to protected zones (Belikov and Legare 1996: 9). In many Arctic countries, protected areas are co-managed with indigenous and local peoples, through whom access to resources is maintained and knowledge is shared. Traditional knowledge provided through co-management allows indigenous perspectives to contribute to protected areas management. By maintaining ecological integrity, protected areas can help maintain the spiritual and traditional lifestyles of ArcticIndigenous peoples (Barry and McLennan 2010: 96).

Greenland has a small population with a fragile economy, based essentially on a grant from the Danish Government and on fishery. Declining fishing revenues and the fragility of this industry lead the Greenland Government to search for other economic opportunities, notably the mining of Greenland's rich oil (offshore) and mineral deposits. Greenland has no exploitation of hydrocarbon resources today, but exploration is taking place. There is broad political consensus that measures should be taken to develop the mineral resource sector into one of the mainstays of the economy (Fourth National Report 2004; Ramsar report 2009). The plans by the Greenland Government to establish an aluminium smelter and to extract offshore oil and minerals, such as gold, iron, diamonds, rubies and others, from Greenland's soils are intended to complement the fragile hunting and fishing economies by new industries and to make the country economically more self-sufficient. These plans emerged at a significant moment when the Greenland Government moved (on the National Day of 21 June 2009) from thirty years of "Home Rule" to "Self-Governance", providing it with more autonomy from Denmark (Fourth National Report 2004; Ramsar Report 2009; Hjarsen 2003).

After the 2009 reorganization of the municipalities, most of the Ittoqqortoormiit population feel more peripheral than usual and not only in a geographical sense.

---

grouse, cod, halibut, lump sucker, salmon, red fish, crab, shrimp and molluscs. Many of these species occur in protected areas although the monitoring effort varies with the species;

- Threat monitoring including monitoring the number of expeditions/visitors (East Greenland National.
- Park and other protected areas), monitoring ad hoc visits to at least one protected area, monitoring harvested species in protected areas with quotas (e.g., polar bear, walrus, narwhal, beluga, musk ox and caribou) and monitoring grazing effects at two sites, neither of which are in protected areas; and,
- Local monitoring by non-scientists including patrol-based recording of wildlife by Sirius Sledge Patrol in East Greenland National Park, community-based monitoring of selected species, threats and climate parameters (under development) and a public observation database (under development) where members of the public can report sightings of species, climate observations and observations of other environmental matters (Livingston 2011:20).

This is transposed in terms of a decrease in services, and most of all in moving many of the local competences to Nuuk, the regional capital of Sermersooq (and the capital of Greenland), which all resulted in a lack of local representation and focus point. Nevertheless, the community is trying to find a way through the many changes that have affected this fairly small and peripheral place. The new rules of the national park open the possibility of developing tourism activities in the protected area. This could revitalize the community, and create some additional tourism jobs like guiding, sledging, and boat rides and give way to new seasonal tourism business that will bring benefits and enhancement to the community of Ittoqqortoormiit.

References

Barry, T., & McLennan, D. (2010). *Changes in protected areas* (Indicator 21, Arctic biodiversity trends). Akureyri: CAFF.

Belikov, S., & Legare, G. (1996). *National mechanisms and principles for protected areas in the Arctic countries* (Habitat Conservation Report No. 3). Akureyri: CAFF.

Boertmann, D., Mosbech, A., Schiedek, D., & Johansen, J. (Eds.). (2009). *The Western Greenland Sea. A preliminary strategic environmental impact assessment of hydrocarbon activities in the KANUMAS East area* (Technical Report No. 719). Aarhus: NERI.

Bugge Jensen D., & Christensen K. D. (2003). *The biodiversity of Greenland – A country study*. Technical Report No. 55. December, Nuuk: Pinngortitaleriffik, Grønlands Naturinstitut.

CAFF. (1996). *Proposed protected areas in the circumpolar arctic* (Habitat Conservation Report nr. 2). Trondheim: Directorate for Nature Management. http://caff.is/monitoring-series/19-protected-areas-monitoring/165-proposed-protected-areas-in-the-circumpolar-arctic-1996

Danmarks Statistik. (2009). Statistisk Årbog 2009. Danmarks Statistik, København. http://www.dst.dk/Site/Dst/Udgivelser/GetPubFile.aspx?id=13457&sid=saa2009

Danmarks Statistik. (2010). Statistisk Årbog 2010. Danmarks Statistik, København. www.dst.dk/pukora/epub/upload/15197/saa2010.pdf

Danmarks Statistik. (2011). Statistisk Årbog 2011. Danmarks Statistik, København. http://www.dst.dk/da/Statistik/Publikationer/VisPub?cid=16217

Due, R., & Ingerslev, T. (Eds.). (2000). Naturbeskyttelsi i Grønland. Teknisk Rapport nr. 29, Nature protection in Greenland, Technical Report 29, Nuuk: Grønlands Naturinstitut, Greenland Nature Institute.

Egevang, C., & Boertmann, D. (2001). *The Ramsar sites of Disko and West Greenland* (NERI Technical Report nr. 368). Roskilde: Ministry of Environment and Energy, National Environment Research Institute, Department of Arctic Environment.

Frazier, S. (1999). *Ramsar sites overview*. Wageningen: Wetlands International.

Geografi, Klima og Natur. (2013). Nuuk: Statistisk Årbog .

Government of Greenland. (2004). *Fourth National report on the implementation of the convention of biological diversity of Greenland*. Nuuk: The Ministry of Domestic Affairs, Nature and Environment.

Hjarsen, T. (2003). *Greenland's international obligations – A report on Greenland's fulfilment of international conventions and agreements on nature protection, Species conservation and Wildlife management*. Denmark: WWF.

Jepsen, P. U., Søgaard, B., Ragborg, A. G., & Møller, H. S. (1993). *Danish report 1993 on convention on Wetlands of international importance especially as waterfowl habitat in Denmark and Greenland (The Ramsar Convention)*. Copenhagen: Ministry of Environment and Energy, The National Forest and Nature Agency.

Kingdom of Denmark: Strategy for the Arctic 2011–2020. (2011). The Ministry of Foreign Affairs of Denmark, Department of Foreign Affairs of Greenland and the Ministry of Foreign Affairs of the Faeroe Islands.
Koivurova T. (2009, June). Governance of protected areas in the Arctic. *Utrecht Law Review*, p. 45, 5(1).
Livingston, D., et al. (2011). *Circumpolar protected areas monitoring. Arctic protected areas monitoring scheme background paper* (CAFF International Secretariat, CAFF Monitoring Series Report Nr. 5).
Ministry of Foreign Affairs. (2011). *Kingdom of Denmark strategy for the Arctic 2011–2020*. Copenhagen: Ministry of Foreign Affairs.
Ministry of Housing, Nature and Environment. Nuuk. Retrived from http://naalakkersuisut.gl/da/Naalakkersuisut/Departementer/Boliger-Natur-og-Miljoe/Naturafdelingen/Nationalparken-i-Nord-og-Oestgroenland
National Park/Man and Biosphere Reserve – Workshop. (2010, May20–21). Retrived from http://naalakkersuisut.gl/da/Naalakkersuisut/Departementer/Boliger-Natur-og-Miljoe/Naturafdelingen/Nationalparken-i-Nord-og-Oestgroenland/Workshop-2010
Nickels, S., Milne, S., Wenzel, G. (1991). Inuit perceptions of tourism development: The case of Clyde River, Baffin Island, N.W.T. *Etudes/Inuits/Studies, 15*(1):157–169.
Nordic Council of Ministers. (2006). *Nature and cultural environments in the Arctic, Greenland, Iceland and Svalbard* (TemaNord). Copenhagen: Nordic Council of Ministers.
Olsen, R. H. (2010). *Interests of the municipality of Kommuneqarfik Sermersooq*. Nuuk: Naalakkersuisut, The Self Government of Greenland.
Pagnan, J., Legare, G., et al. (2004). *Conservation of Arctic Flora and Fauna program, values of Arctic protected areas: A summary of "Protected areas of the Arctic: Conserving a full range of values"*. Akureyri: CAFF.
Ramsar Convention Secretariat. (2010). *Ramsar handbooks for the wise use of wetlands* (4th ed.). Gland: Ramsar Convention Secretariat.
Salathe, T. (2009). *Ramsar Advisory Mission No. 61: Greenland*. Gland: Ramsar Convention Secretariat.
Statistics Greenland. (2013). *Greenland in figures*. Nuuk: Statistics Greenland (in Danish). http://www.stat.gl/dialog/main.asp?lang=da&version=201302&sc=SA&subthemecode=o3&colcode=o
Statistics Greenland – Nature. (2013). Nuuk: Statistics Greenland.
Statistics Greenland. (2009). *Statistical yearbook*. Nuuk: Statistics Greenland.
Tommasini, D. (2009). *Community-base tourism as option for concrete, viable development in peripheral, remote places: Ittoqqortoormiit, East Greenland*. Report from field work, delivered to the Ministry of Education and Research.
Tommasini, D. (2011). Tourism experiences in the Peripheral North. Case studies from Greenland. Inussuk Vol.2, Nuuk: Government of Greenland, Ministry of Education and Research.

Laws and Regulations

Act on Greenland Self-Government, Act no. 473 of 12 June 2009.
Executive Order no. 7 of 17 June 1992 from the Greenland Home Rule Authority concerning the National Park in North and East Greenland, as amended by Executive Order no. 16 of 5 October 1999. Act of Landsting no. 11 of November 12th 1980 concerning Nature Preservation, The Greenland Home Rule Authority, 17 June 1992, Amended by Executive Order no. 16 of 5 October 1999, taking effect at 1 December 1999.
Greenland Home Rule Executive Order Concerning Preservation of Arnangarnup Qoorua, Maniitsoq Municipality, West Greenland. Act of Landsting no. 11 of November 12th 1980 concerning Nature Preservation, Greenland Home Rule, Nuuk, October 20th 1989.
Greenland Parliament Act no. 7 of December 2009 on mineral resources and mineral resources activities (The Mineral Resources Act).

Conflicting Understandings in Polar Bear Co-management in the Inuit Nunangat: Enacting Inuit Knowledge and Identity

Stéphanie Vaudry

Abstract The co-management of polar bears between scientists and the Inuit in Nunavut has been fraught with tension. This chapter explores the Inuit's perspective by highlighting where the bear fits within Inuit cosmology and how it influences their relationships with the animal, with respect to hunting. Since 2005, environmentalists have tried to ban polar bear hunting on an international scale and to get the animal put on the list of species threatened with extinction. This has had a major impact on already fragile northern economies, as it discourages sport hunting, which many Inuit count on for needed income. After analyzing the data, it appears that sport hunting has positive effects on Inuit communities: it provides economic and material resources all throughout the year, but also allows for the reaffirmation of Inuit identity and the transmission of Inuit knowledge to younger generations.

Keywords Co-management conflicts • Inuit Qaujimajatuqangit • Inuit Knowledge • Polar bears sport hunting • Nunavut

1 Introduction

> Wildlife biologists make hunters unhappy. They make regulations and apply them to us without our consultation. These are policies without thought that make our lives difficult (in Kunuk 2010).

The hunting of polar bears (Ursus maritimus) in the Canadian Arctic has been subject to controversy since the 2005 call from the Center for Biological Diversity (CBD) to place the animal on the list of threatened species. This reaction came about following the Government of Nunavut's initiative to increase quotas in January 2005 (see Dowsley and Wenzel 2008). From that time on, there have been several advertisements, such as that of Coca-Cola in partnership with the World

S. Vaudry (✉)
Department of Anthropology, Université Laval, Québec, QC G1V 0A6, Canada
e-mail: stephanie.vaudry-gauthier.1@ulaval.ca

T.M. Herrmann, T. Martin (eds.), *Indigenous Peoples' Governance of Land and Protected Territories in the Arctic*, DOI 10.1007/978-3-319-25035-9_8

Wildlife Fund (WWF), that attempt to raise awareness about this cause. In the wake of much lobbying and campaigns whose claims found support in several biologists' publications, the United States placed the polar bear on the US Endangered Species Act (ESA) in 2008, and thus began regulating the importation of products derived from the bear as well as advocating for the same to be done on an international scale, which has had a direct impact on revenues from sport hunting in the North. Moreover, the only Inuit counterarguments that have been recognized are those containing scientific data. However, far from backing down from these allegations, Inuit peoples, mainly in the framework of the Government of Nunavut, have refuted the "facts" pointing out that there are more bears than before, and that the studies are based on non-exhaustive data and detached from reality. So they decided to kiumajut, to talk back.

There seems, therefore, to be some contention in terms of how to approach this issue. In this sense, the question I have asked myself is why, in the first place, there is this conflict around the management of polar bears, which has, in turn, lead me to wonder in what way the Inuit's vision is different from that of the scientists, what the practical implications of these differences are, and finally whether or not sport hunting is a new development or inscribed in a continuity with this vision and how. I argue that the dispute lies in the difficulty to acknowledge the concrete implications of Inuit "universe of meanings" (see Gagné 2013: 12 about his concept) with regards to the hunting of polar bears. The current struggle of certain NGOs, scientists and elected officials to classify the animal as an endangered species stems from, as we will further examine, is an assault on the Inuit relationality, that is, one's sense of interconnectedness with human and nonhuman entities (See McGrath 2011; Wilson 2008). This type of policy can have undesirable consequences on the future of Inuit identity and the survival of a people that are intimately connected to hunting,[1] which is carried out in mutual respect and is a way in which their identity is maintained and passed down. This study places Inuit concepts regarding the management of polar bears front and centre, an approach that is all too often overlooked in the pursuit of more technical aspects of research.

In order to adequately frame the matter at hand, I will first present a brief context of the conflict around the management of polar bears, and will then examine the concepts of Inuit Qaujimajatuqangit and actor-network in the hopes of understanding the importance of this animal for the Inuit people, which I will later apply to the question of hunting.

[1]Martin (2003: 96, 115) points out that the hunt, aside from its contribution to food and economic autonomy, brings in much needed revenue to Inuit communities.

2 Why Ban the Hunt?

2.1 The Current Juncture

The image of the polar bear has been increasingly associated with climate change and, by extension, irresponsible behaviour towards the environment: "You can call it an icon, a flagship or a canary in a coalmine. [...] This is an indicator of some impact of something that humanity is doing, something that is going on, and it's expressed in a simple way by the polar bear" (Peter Ewins, director of the WWF Toronto, quoted in O'Neill 2007: 18). Several NGOs, including the WWF and the CBD have hoped to strike a cord with people and get them to alter their behaviour. Supported by western scientific research, these claims highlight the chasm between supposed "rational" knowledge and that of the Inuit regarding the management of polar bears, which may be putting northern Inuit communities at risk.

Recent studies have surfaced on the quantity of polar bears in the Artic Circle that raise concerns about their future due to a purported decline. Several reports have also been published on three of the nineteen ursus maritimus populations, indicating that their health has been deteriorating, while estimating that this finding can also be applied to other populations and that the situation will get progressively worse (Derocher et al. 2004). These documents shed light on the negative impacts of melting ice on polar bears populations, which include: general health and reproductive issue in females, higher infant mortality, scavenging and even "cannibalistic" activity (Derocher et al. 2004: 170). Polar bears have, therefore, been facing a series of challenges to stay fed and avoid starvation. This is because the layer of ice allowing the bears to hunt seals has not been as solid and not lasted as long, thus inhibiting them from building up sufficient reserves of fat for the remainder of the year (Derocher et al. 2004: 166). In this sense, melting ice reduces access to food, which, according to several western scientific reports, is due to and being worsened by climate change. It is now estimated that by 2050, polar bear populations will reach critical levels of malnutrition and famine (Derocher et al. 2004: 164), making extinction seem imminent.

These scientific "discoveries" are based on satellite photos (Taylor et al. 2008) and quantitative data from the Arctic on certain populations, mainly in Hudson Bay where the population has decreased by 25 % (O'Neill 2007). As mandated by the 1973 accord signed by countries[2] with polar bears in their territory, a census is taken every fifteen years (Wong 2010). With the first studies only dating back to the 1970s, there are no archives, beyond Inuit oral history, that can report on the situation before that time.

According to several Inuit representatives and elders, these western scientists are mistaken on many levels. They criticize the many inadequacies of the studies in terms of the preconceived notions about the bears' habits, the blatant denial of their

[2]The signatories included Canada, Denmark, the United States, Norway (Greenland), and the Soviet Union (in 1974).

adaptability and, above all, the fact that the lives of the polar bears seem more important to them than those of the Inuit. These events have been reminiscent of the impact (destruction of local economies, and the undermining of Inuit values and worldviews) that the 1983 European Union boycott of products derived from the seal[3] hunt had on their lives (Tester and Irniq 2008; Wenzel 1989).

First of all, they take issue with the way in which western scientific research has been conducted, evidenced in their inadequacies and detachment from reality. In actuality, these studies on the transformation of polar bear populations are not exhaustive, as they were carried out on only three out of the nineteen populations (Dowsley 2009a). Different populations have very different patterns of behaviour in terms of the ground they cover, the way they socialize and the spaces they inhabit (Randa 1986: 60–61). In addition, these studies were conducted during the summer and have hardly considered their intermingling and distance traveled. The Inuit equally question the time elapsed between different studies, which do not adequately take into account the transformation of populations. As previously mentioned, data was collected at the start of the 1970s, to be later repeated every 15 years. Other studies have also taken place, but they were limited in scope (Dowsley 2009a; Tyrrell 2006). Therefore, said research does not offer a representative look into the animal's behaviour. What is more, several Inuit believe that the biologists have had a detrimental impact on the bears, as the tracking collars they attach to them inhibit their ability to hunt seals by making it difficult to straighten their necks to look down into holes in the ice (Kunuk 2010). They also emphasize the fact that the bears have a delicate sense of hearing, which is negatively affected by the coming and going of helicopters overhead. All in all, the leaner bears appear to be those that came into contact with the researchers (Kunuk 2010).

According to Inuit universe of meanings, bears are intelligent beings able to adapt to changes in their environment (Kunuk 2010; Tyrrell 2006). If the hunting of seals gets to be difficult in one area, the bears will compensate by going elsewhere. Certain elders, in fact, have described how there are abundant bird nests in crevasses for the bears (O'Neill 2007), while others have mentioned that they head farther up north in pursuit of the seals (Tyrrell 2006) like they have always done[4] (Randa 1986: 46). They have similarly pointed out that their increased presence near villages is a sign of their adaptability, as they know very well that they will find the Inuit's left-over food: "There seems to be more damage, but you have to take into consideration that we're leaving more of our stuff on the land than fifteen years ago. But if you leave meat catches they are pretty much guaranteed to be gone" (quoted in Dowsley 2007: 63). Several families have stated that the bears take the meat that they leave outside, even if it is buried under rocks; they adjust. This phenomenon is no surprise to the Inuit, as the movement of bear populations was also linked to their own

[3]A polar bear kills on average six seals per month, mainly young seals, to sustain itself; significantly less than an Inuk. The bear thus poses a greater threat to seal populations; nevertheless, environmentalists direct their attention towards the Inuit (Randa 1986: 83).

[4]Polar bears could once be found as far south as the Saint Lawrence Valley (Randa 1986).

movements when hunting, in that the bears would devour the carcasses they left on the trail (Randa 1986: 51).

As stated by several hunters, the polar bears' quest for food is not as bad as the scientific predictions would suggest. However, the scientists and Inuit do agree on one point: melting ice has an impact on the thickness of the ice sheet. Inuit have reported that with the melting of the icebergs the edge of the ice has come closer to the land, as the icebergs are no longer holding it in place (Dowsley 2007: 66). However, the Inuit do not necessarily share the interpretation on how this impacts the bears. Certain elders have claimed that the thinning of the ice does not have an effect on polar bears ability to hunt given that they would benefit from summer in which it is easier for them to catch seals in the water (Kunuk 2010). The bears would hunt seals closer to the water's edge and not where the ice is thick, as such, climate change would increase their presence on the coast (Tyrrell 2006). Once again, the Inuit demonstrate the general lack of knowledge on this animal's habits.

Nevertheless, what most bothers the Inuit is that the threat of the bear's disappearance seems more important to scientists and elected officials than human life (Clark et al. 2008). In fact, several Inuit representatives have made clear that polar bears, found in ever increasing numbers in villages, have posed an escalating danger to their communities (Berkes and Armitage 2010), as well as the fact that climate change has jeopardized Inuit lives in general (Kunuk 2010). According to numerous residents, there have been more bears than before, and in closer proximity to communities, obliging the Inuit to modify their habits and become more vigilant, both in their villages and when out hunting, by alternating techniques and routes (Kunuk 2010; Tyrrell 2006). They have to defend themselves. On a spiritual level, to not do so would show a lack of respect to the animal and risk further reprisals (Tyrrell 2006). Furthermore, strict quotas could augment the risks they already confront.

Faced with differing conceptions on the management of polar bears, co-management has come into fertile terrain as a space in which compromises can be found. However, in reality it has been quite the contrary. The rise of Traditional Ecological Knowledge (TEK) has lead to the development of systems of management between western scientists and Indigenous peoples, regarding which various researchers have mentioned the case of polar bears in northern Canada as a good example of this method (Clark et al. 2008; Dowsley 2009a; Martin 2009; Wenzel 1999).

Starting in the 1980s, the Government of Yukon implemented a system to integrate Indigenous knowledge in biodiversity management report submissions (Nadasdy 1999: 14). From 1999 on, the Nunavut Wildlife Management Board has been the body for the co-management of wildlife between the Government of Nunavut and Fisheries and Oceans Canada (Armitage et al. 2011: 997), in which Inuit can submit *recommendations* relative to their knowledge of the land. In other words, they only act as counsellors; they have no decision-making power.

Polar bear co-management boards espouse a western scientific methodology in which data is "extracted" from elders in order to better understand an animal in its setting (Nadasdy 1999). During the interviews, scientists ask mainly quantitative, i.e. measurable, questions on a given animal population. Nadasdy (1999) recounts that researchers have the "frustrating" tendency to compartmentalize and filter Indigenous knowledge. Likewise, several elders have stated that when they delve into more qualitative aspects, namely when explicating the context in which the bear lives and how it integrates with the Inuit, the researchers seem to look down upon this information (Nadasdy 1999; Tester and Irniq 2008). They feel their contribution is not valued or considered important to the decision making process, about which Nadasdy (1999) adds that even if the board's recommendations were to go directly to the minister's office, a team of white scientists would go back and filter the data, once again glossing over the qualitative significance[5]. In the literature, others have brought attention to the fact that there are no decisions made, or reports published, in which TEK is the sole source – with the exception of Nunavut, which, in 2004, pushed for Inuit Qaujimajatuqangit (IQ) as a primary source (see Laugrand and Oosten 2002; Peters 2003: 56). IQ is similar, in principle, to TEK, except that it entails a more dynamic process of production, sharing and acquisition of knowledge and that it draws on Inuit cosmology. In this sense, TEK would be more like a set of insights for scientific research or a sort of confirmation of data already collected (Peters 2003: 55), and not a determining factor in the decision making of land management in which ultimately it is scientific knowledge that validates results (Nadasdy 2003: 375; Peters 2003: 54).

It is also important to add that the co-management board meetings are held in urban centres, at the offices where the scientists work or where government buildings are located. This means that informants must travel from their land in order to participate. Consequently, there is a distorted sense of time and space in the accounts offered by indigenous people on their knowledge and relations to the land (Nadasdy 2003: 375). The fact is that the accounts are given in a space removed from their territorial context, which makes the connection with Indigenous knowledge and the understanding of its pertinence more difficult. In that respect, Nadasdy (2003: 374) specifies that he was only able to understand the importance of TEK when in the field. It is also important to add that participants are not accustomed to the offices in which the interviews are held or the format of the meetings. In these conditions they do not always feel comfortable speaking about their experiences with and knowledge of the land (Tester and Irniq 2008: 57). The information the scientists extract can be thus left to open interpretation.

Ultimately, co-management bodies are intermediaries between the state and Indigenous nations and aim to reinforce the power of the state over these nations' lands (Nadasdy 2005: 223–225). The three elements presented are not necessarily constitutive of the general uneasiness with co-management, but rather indicative of the real will to exercise political power in the decision-making, and increasingly, in the political autonomy so vehemently demanded by Indigenous peoples. Through multiple consultations and requests for data compilation, co-management has lead

[5]A master research regarding the Huron-Wendat territorial management and claims with the Government of Québec recently shows that it is more the bureaucratic structure of the state and its informal policies than the data itself that complicates the process, since this Indigenous nation produces reports made by Indigenous scientists (François-Xavier Cyr, personal communication, 2015).

to a bureaucratization of Indigenous peoples (Nadasdy 2005: 224), by impelling them into a system of management of their own ancestral lands whose parameters are dictated by the government (Nadasdy 2005: 226–227). This becomes evident in meetings where the emphasis is placed on the technical aspects of the land, that is, by gauging applicable knowledge and determining whether or not to manage "the resources" differently. The fact that Indigenous people themselves might be the best managers is never discussed, or that they alone would be able to decide on how the land should be administrated (Nadasdy 2005: 228). Co-management is, therefore, a tool that serves to better control Indigenous peoples by avoiding the more political questions that lead to real change and could potentially call into question the sovereignty of the settler colonial state in these lands (Nadasdy 2003: 378).

2.2 A Source of the Problem

The universe of meanings attached to Inuit conceptions of reality correspond to, in a scientific and western sense, non-traditional ways of seeing things and adapting to everyday life. The implementation of IQ by the Government of Nunavut in 2004 has laid to the foundation of a better understanding and transmitting of the Inuit knowledge, by reaffirming and revitalizing their identity. The diverse interactions of the Inuit world, be they spiritual or practical, run counter to mundane standards. Latour (2006), via his actor-network theory, helps us better understand the importance of these relationships and how nonhumans can act as mediators in diverse situations. For the Inuit, interactions take place not only amongst themselves, but amongst objects and animals as well.

As previously mentioned, Traditional Ecological Knowledge is a tool for the integration of Indigenous knowledge into a "western" scientific research (Nadasdy 1999) and the reaffirmation of the power of the state in their land (Nadasdy 2005: 223–225). Conversely, Inuit Qaujimajatuqangit[6] (IQ), encompasses traditional ecological knowledge (TEK) as well as non-empirical cultural aspects that facilitate an understanding of the natural world (Dowsley 2007: 54). Fundamentally, Inuit knowledge comes from the elders' memory (Martin 2009) and could be defined as: "A set of teachings on practical truisms about society, human nature and experience passed on orally (traditionally) from one generation to the next… It is holistic, dynamic and cumulative in its approach to knowledge, teaching and learning" (Jaypetee Arnakak quoted in Martin 2009). However, it is equally a present-day body of knowledge: "the Inuit way of doing things, and includes the past, present and future knowledge of Inuit society" (Bell 2002: 3 quoted by Tester and Irniq 2008: 49). IQ is, therefore, dynamic and not confined to the past, but rather allows youth to integrate and innovate the knowledge and thus negotiate their experiences with the worlds in presence.

[6] *Inuit Quaujimajatuqangit* can be translated as "traditional Inuit knowledge," a product of *quaujimajaq*, or 'that which is known.'

IQ is the product of Inuit struggle to preserve and revitalize their culture and identity by way of *resistance* to governmental decisions that tend to be made on the basis of scientific advice that directly goes against their way of subsistence and identity (Nadasdy 2005; Tester and Irniq 2008). IQ is ipso facto tied to Inuit identity, insofar as it recognizes the existence and the value of Inuit knowledge by giving it an active role in the bolstering and safeguarding of their traditions (Laugrand et al. 2000; Martin 2009). It is a site of resistance in which the Inuit are the protagonists (Laugrand and Oosten 2002; Tester and Irniq 2008), as it reconciles different approaches to the global on a local level (Nadasdy 1999). It is also a political tool offering a proven effectiveness in social struggle (Martin 2009), because of which the Inuit have taken a more protagonist role in decision making processes; IQ poses a challenge to hierarchies of knowledge. As such, the Inuit have been increasingly recognized as active agents, having placed at the forefront the different transformations they experience, which underscore their creativity and adaptability in terms of change (Laugrand and Oosten 2002). This reflects mainly in their will to bring the two types of knowledge together and invent something new that allows them to take the best of these universes of meanings and navigate through them (Stuckenberger 2010: 7).

According to Usher (2000: 186), Indigenous ecological knowledge by definition consists of four main categories:

- Rational/factual knowledge about the environment;
- Factual knowledge about past and present uses of the environment;
- Culturally based value statements about how things should be and what is fitting/proper to do;
- Culturally based cosmology by which information derived from observation and experience, and instruction is organized to provide explanations and guidance.

This definition is interesting because it offers an understanding of Inuit knowledge as dynamic and links it to non-western worldviews. It also locates different intertwining dimensions of the Inuit way of life. Tester and Irniq (2008) show that, in co-management initiatives, it is often the last two aspects that are usually disregarded due to their qualitative character. The fourth aspect is even more ignored since it is a product of spirituality and is therefore incomprehensible, i.e. senseless, from the rational point of view of western science. This dimension intersects Nadasdy's (1999) research on co-management in which he had pointed out elements seemingly incongruous to Indigenous thought, i.e. the compartmentalization or distillation of knowledge by eliminating all qualitative elements and keepings all that is measurable. According to Nadasdy, it goes well beyond difference indicating, rather, two opposing models: "That is rather being holistic, oral, qualitative, and intuitive, TEK artefacts tend to be categorized, written, quantitative, and analytical" (1999: 9). Moreover, Tester and Irniq (2008) explain in their text that when confronted with more spiritual questions white researchers invariably let out a sigh during interviews. The problem of co-management can be located, in part, in this element, however it also shows Inuit people's will to superimpose other ways of doing.

The actor-network theory, advanced by Latour (2006), emphasizes the importance of taking into account human and nonhuman elements together as actors in a given situation. Latour suggests that one let associations happen freely between humans and objects. Inanimate objects, or animals, may have an important meaning for the actor and influence his or her actions. In actuality, for Latour, they all act as mediators; they transform, translate, distort and modify the meanings and elements they are supposed to transport, given that they interact with humans. Humans can then respect non-humans. As I will examine in the following section, bears are respected by the Inuit with whom they interact. In that sense, the approach outlined by Latour suggests that there are multiple rapports amongst humans and nonhumans alike and that they can have all different types of associations at many different levels. Recently, scholars, such as McGrath (2011) and Wilson (2008), have identified this "relationality" as being the core of Indigenous peoples' worldviews – saying that Indigenous peoples are relational in their sense of being in and seeing the world as well as relational accountant in their everyday thoughts, attitudes and actions to all past, present and future human and non human beings. This framework of analysis allows for a better understanding of IQ, as well as for western scientists' lack of understanding of the Inuit's position regarding the hunting of polar bears.

3 Bear and Inuk, a Story of Equals

The research carried out by these scientists is said to be detached from reality and insufficient by Inuit. This problem fundamentally resides, according to Schmidt and Dowsley (2010), in the fact that the scientists perceive the polar bear as a passive being vis-à-vis the activities taking place on the land and concentrate their efforts on the Inuit's "agency." According to the Inuit, polar bears reflect on their actions; they can feel what others think and intend to do. They are agents. This idea breaks with the "western" logic of the commons – thought of as collective property where the plants and animals are resources – which poses a problem in the management of lands and, specifically in this case, hunting.

3.1 The Bear as a Living, Thinking Being

Many Inuit have highlighted the resemblances that have come to define Inuit and polar bear, as well as the sense of closeness they feel. According to the Inuit, the bears are powerful and intelligent beings (Dowsley and Wenzel 2008; Schmidt and Dowsley 2010). Not only through their rapid adaptation to contemporary challenges, but also their creative way of negotiating it, the Inuit have interwoven the different ways of doing through which they have cleared a path (Berkes and Armitage 2010; Martin 2003). In terms of polar bears, the Inuit have alluded to these same qualities that they seem to be using to confront various issues when hunting and adapting to climate change in general (Berkes and Armitage 2010). For example, as we saw above, since seal populations have gone farther north and the edge of the ice sheets are now closer to land, the bears have learned to follow them and hunt closer to the banks.

Table 1 Hunting techniques commonly employed by both polar bears and Inuit (Randa 1986: 149)

Polar bear *nanuq*	Inuk
Catching seal in their breathing hole, or *aglu*	Harpoon
Catching (young) seal in their den	Hook, harpoon
Catching seal lying close to their *aglu*	Harpoon, rifle – approach masked behind a white screen
Catching seal in the water, near the edge of the ice	Harpoon, rifle
Catching seal in the water (swimming)	kayak

In the past, the Inuit would build igloos during the hunting trips in the winter and tents in the summer; today they use tents when hunting and live in small houses. Polar bears also build their own shelter by digging in snow to make a den in order to protect themselves from inclement weather (Randa 1986).

The physical resemblance is also highlighted; just like the Inuk, the bear is a "biped." In fact, according to several Inuit accounts, it seems that, underneath the fur, its body looks very much like that of a human being (Schmidt and Dowsley 2010).

The Inuit have also pointed out similarities between their hunting techniques, given that they are the two main predators in the Arctic (see Table 1). For example, polar bears will cover all but one breathing hole with snow in a given location to try to catch the seal when it comes up for air. This technique is called aglu (Randa 1986: 100). During the summer, the bear tries to cover its black nose with its paw or hide behind other obstacles to blend into its white surroundings and catch the seal on the ice (Randa 1986: 102).

The polar bear's hunting techniques have clearly inspired Inuit hunters, as they use those same tactics themselves. In this sense, the Inuit perceive the bear as their equal on many levels, insofar that it is a being capable of acting on its environment and can find solutions in order to adapt; it is not only a victim.

The Inuit hold an intricate relationship with the bears that cannot be merely reduced to two different species. In Inuit carvings, one finds scenes in which the bear and the Inuit interact with each other, as well as with other animals (see Povungnituk). The bear can be seen in numerous artistic depictions, in which it is presented as a powerful being that the Inuit fear. It attacks humans and goes after the biggest animals in the north, such as walruses. Hence, in order to not fall victim, the Inuit must use a variety of tactics and stick together, which is where the hunting of polar bears comes in. For this, dogs are essential as they distract, bite and attack the bear, buying time for the Inuit hunters to get into position and neutralize it: "Dogs also alerted hunters on the ice to the presence of a bear. They chased it, surrounded it, and kept it in the bay until the hunter arrived" (Bennett and Rowley 2004: 285). This collaboration is crucial to the polar bear hunt, as it is an intelligent animal that anticipates attack.

3.2 The Bear as a Sentient Being

In the Unikkaaqtuat,[7] polar bears, along with many other animals, abound. The Inuit often transform into animals, and animals into people. Actually, according to Inuit stories, animals and humans were not distinct beings: "All animals could turn into people, according to what we were told. They turned into people a long time ago before there was Christianity" (Lucassie Nutaraaluk quoted in Martin 2009: 194), and these transformations explicate several elements of the Inuit's lived experience. In that sense, the polar bear has a special status; it is perceived as their equal due to its strength and intelligence. In keeping with one of the central points of this people's cosmology: "We find the bear's presence in the depths of the ocean as well as in the heavens, in all the realms of the sea, land and air, in associations with life and the powers of the greatest spirits as well as the weakest humans (the orphans)" (Saladin d'Anglure 1980: 183). The Inuit explain and attribute many of the events in their daily lives to the bears. As such, they have a considerable amount of respect for them.

In several cases, the Inuit have alluded to the bears' acute ability to feel people's intentions. This keeps them on guard as to what they may say or think about the bears (Dowsley and Wenzel 2008; Tyrrell 2006). It is often recounted that a bear attacked so-and-so because he or she did not respect them:

> I don't really know [why a bear might attacks someone]. Maybe it is that we are not supposed to say bad things about polar bears. When a man's property is damaged he might get mad. We are told polar bears have minds like humans. The man might threaten to kill that polar bear. The polar bear also knows there are seasons when humans can't kill polar bears and if a man kills one out of season the polar bears might get mad (quoted in Dowsley and Wenzel 2008: 186).

Thus, the bears are supposed to be used to today's conditions and also function via the system of quotas and hunting seasons, which is where one of the facets of their mutual respect lies. Certain elders have also stated that the damaged inflicted on property in their villages is due to a lack of respect for the bears when they offer themselves up and the people refuse to hunt them because of the quotas. For this reason, practices stemming from the compliance with the restrictive and non-flexible quotas may be putting the lives of several Inuit in danger (Berkes and Armitage 2010).

Undoubtedly, the relationships between polar bears and the Inuit are quite distinctive, as they do not reflect a logic of dominant-dominated; quite the contrary. In accordance with the elements of Inuit universe of meanings herein presented, the Inuit and the bear are very similar beings and interact holistically. The fact that the animal is very much respected by the Inuit also implies that actions that may negatively affect it are denounced in the hopes of quelling any disrespectful behaviour towards it. The polar bear's prominence as an actor in the hunt reflects the Inuit's perception of it, as well as the reasons why they do not easily give in to the scientists. From this perspective, the hunt appears to be a significant aspect in

[7] Age-old stories, or legends, in Inuktitut.

understanding how this special relationship plays out in practice and how Inuit identity is being revitalized.

4 Hunting as a Form of Identity Revitalization

Hunting is a reflection of the multiplicity of relationships that the Inuit maintain with the bears, and with themselves, while being intrinsically tied to Inuit identity. Specifically, subsistence hunting is a source of reaffirming the relationship between these two actors as this activity keeps both Inuit and polar bears in contact. It also allows for the transmission of knowledge associated with the hunt, which includes technical skills as well as ways of perceiving things. Also, sport hunting makes it possible for the Inuit to continuously engage in their cultural activities, since it enables them to finance their expeditions and thus refresh their knowledge, in spite of constraints imposed by the capitalist economy. Hunting continues to be a valued activity, not only on a socio-cultural level, but on an economic and subsistence level as well.

4.1 Subsistence Hunting

Hunting polar bears, for the Inuit, is tied to notions of respect for the animal, whose foundations are grounded in Inuit cosmology. Its practice has the effect of reaffirming and passing down the multiple practical forms of knowledge affiliated with older generations and, consequently, the perpetuation of the relationships that the Inuit maintain with the bears as well as with other Inuit communities. As such, the hunt has a direct correlation to Inuit identity, as it encompasses the four dimensions of their worldview found in IQ.

The Inuit polar bear hunt rests on three fundamental principles. It must be done: (1) Out of necessity; (2) In a spirit of sharing and; (3) When there is a common understanding between bear and hunter (Randa 1986). In this sense, it is done in a way that is respectful to the bear. Moreover, these dimensions are directly reflected in Inuit legends, such as "The Woman Who Had a Bear as a Foster-Son" (Rasmussen and Worster 1921); both bear and man must be mutually respected. Today, the hunt is regulated by quotas that are locally drawn at random and distributed for a period of 24–48 h to members of the HTOs (Hunter and Tapper Organizations) (Dowsley 2005). This manner of proceeding has altered the ways in which people relate to each other, to the bears and to the land, but at the same time, has also worked to maintain them (Dowsley 2009b).

Subsistence hunting has continuously been one of the main activities in northern Canada. In part, the Inuit owe their food and material self-sufficiency to it, in principle, hunting for what they need, and no more. Today, the polar bear hunt is even more revered then before, due to the limits placed by the international community on the hunting and selling of other animal species in the north, such as seals and narwhals. However, they continue to persist in spite on the threat of its banning, hunting for the meat with which they feed the people close to them, while also benefiting from the modest earnings of the sale of its hide and other artisanal creations (Martin 2003: 111; Randa 1986). It is done out of necessity, but it is also a way of maintaining and innovating practices beyond just merely catching the bear.

Sharing the benefits from the hunt takes place both within, as well as between communities, amongst extended Inuit families via the Maussian concept of "giving-receiving-reciprocating" (1983 in Martin 2003: 121). Different types of community infrastructure have been developed in order to ensure access to meat, in spite of the constraints imposed by the capitalist economy. In Nunavik, community freezers have been installed in which hunters can share part of their catch – 13 % on average – in exchange for a small sum[8] to help offset the costs incurred while hunting (Martin 2003: 102–103). All community members have access to this food and, like the hunters, can count on it in times of need (Martin 2003: 124–125). Aside from the intermediary position played by the freezers between giver and receiver, the redistribution of products resulting from the hunt also takes place traditionally, in part, thanks to the financial aid that hunters receive (Martin 2003: 125–126). This contemporary infrastructure makes it possible to maintain existing relationships through their diverse rituals steeped in sharing, be they get-togethers, celebrations or the aforementioned community freezers (Martin 2003; Randa 1986).

Having the bear's consent to be captured is one of the most difficult aspects to understand from a Western point of view, as it is directly tied to Inuit cosmology and praxis. According to several elders, the bears get close to the Inuit in order to offer themselves up, because they know they are good people in need of their meat, which they will then redistribute to the rest of the community (Schmidt and Dowsley 2010). As such, the bears choose their hunter after having observed them for quite some time, which is why it is important for Inuit to have a good attitude from childhood, if they are to eventually become good hunters. The polar bear hunt lends a lot of prestige to the hunter and can actually prove to be a distraction due to the excitement the excursion sparks (Randa 1986: 152). However, some feel that the quota system and drawing have a negative impact on the concept of mutual respect, since the bears are now hunted all at once and not when the opportunity presents itself. Likewise, others believe that, since both the Inuit and polar bears are now accustomed to this way of doing things, it should remain as such as long as attitudes are still reciprocal (Schmidt and Dowsley 2010).

The notion of respect in hunting can be found on many levels, but is intrinsically linked to the attitude a person has with his or her peers as well as with the bears. It is not a situation in which one hunts for money, but rather a healthy and comprehensive relationship between two parties; a notion that is not compatible with the

[8]These sums are awarded by the InuitHunting, Fishing and Trapping Support Program, financed by the Government of Quebec and managed by the Kativik Regional Government. The funds help support hunters to carry out their activities throughout the year (Martin 2003: 104). A similar program also exists in Nunavut (see Martin 2003: 116).

quantitative data researchers are looking to obtain and filter in order to construct precise scenarios. What is actually at play, on the contrary, is a set of beliefs that influence several dimensions of Inuit reality drawn from both their cosmology as well as IQ.

It is through hunting, as it were, that Inuit knowledge is lived and passed down. It is what enables the Inuit to possess environmental knowledge relevant enough to understand the polar bears and their habitat (Freeman and Wenzel 2006). Going out to hunt polar bears is an occasion for elders to teach the young about the different ways of negotiating their coexistence with the land and the animals that call it home. This transmission of knowledge is done by imitation and practice, but also through the unikkaaqtuat. The quota system's tag drawing gives a new group of individuals, such women and inexperienced youngsters, the chance to practice polar bear hunting and, in many cases, elders and more experienced men are invited to participate in these outings. They use their prominence to teach the principles of hunting and what the activity entails, and to guarantee that this knowhow is passed down to future generations. By hunting, important knowledge is transmitted and internalized, which allows learners to better understand the world they live in, as well as their history.

Hunting yields other cultural benefits beyond the practice in itself, insofar as practical and conceptual knowledge on how to prepare hunting trips and make use of different parts of the animal is passed along. The preparation of an expedition is done ritualistically drawing both on ancestral and current knowledge, as Inuit hunters combine newer and traditional technologies (Berkes and Armitage 2010). What is more, the procurement of the animal coincides with the transmission of skills to younger generations relative to the preparation of products made from bear hide, which is usually done through storytelling, equally for posterity (Randa 1986). This offers said generation a more comprehensive understanding of where they come from, and the Inuit world in general, as it also extends to such activities as the preparation of meals and their consumption. In other words, the added benefits from hunting are vast and imbricate several dimensions of Inuit identity in a sort of crossroads in which many activities that are meaningful to their identity come together.

In sum, the polar bear hunt is carried out in a logic quite different from that of capitalism and its priorities by inserting itself in an economy of respect and social reaffirmation between the Inuit people, as well as between the Inuit and the bear. It is a mutual relationship. Nevertheless, the practice of subsistence hunting is threatened by a lack of sufficient time and resources available to Inuit communities to sustain the activity. In this sense, sport hunting could prove to be, in many ways, a solution to ensure its continuity.

4.2 *Sport Hunting*

> We cannot exist purely by making money. If we do not have our environment, we cannot survive. Without food, we cannot live… The rest of the world does not care enough about this (Mary Simon, ex-president of Inuit Tapiriit Kanatami, in Kunuk 2010).

Sport hunting provokes different reactions regarding the matter of respect, or lack thereof, for polar bears. Elders have mitigated this by reminding people that one cannot manipulate the polar bears; it must first agree to the hunt, if not it may not ever come back, or cause physical or material damage to members of the community (Dowsley 2007). Going after the bear for tourists' pleasure can have a serious impact on hunting in the future. Several western environmentalist and animal rights groups, such as the CBD and the WWF, also find this practice cruel and dangerous to the bear's survival. Nevertheless, to hunt polar bear, various dimensions have to be respected: the hunters must be hunting out of necessity and are required to share their catch (Dowsley 2009b). Furthermore, the bear must consent to its capture, to which the Government of Nunavut added several other components: guides must be Inuit, use specific weaponry and not travel in motor vehicles. The sport hunt is carried out in accordance with the bear and the quotas (Dowsley 2010). It is an ecotouristic activity that is very lucrative for the Inuit in northern Canada (Dowsley 2009b); however, it constitutes on average 20 % of the quotas and is intrinsically linked to traditional practices (Dowsley 2010). This allows the Inuit to obtain funds for subsistence hunting and also to transmit knowledge relevant to its realization. Consequently, it is an activity that keeps alive and revitalizes Inuit identity.

4.3 The Need to Hunt

As stated by Dowsley (2010), the Inuit do not necessarily profit from the economic side of the hunt, since the amount of quotas allotted does not generate substantial revenue. Profits from sport hunting are often used, in turn, to finance subsistence hunting, mainly for purchasing equipment. In absence of sport hunting, many Inuit would no longer go on outings, for lack of sufficient funds. This means they would not be able to hunt nor reap the benefits, in terms of food or material for artisanal products. Today, hunting is a costly endeavour, and potential revenues would not cover the expense of supplies. However, due to the proportion of quotas (approximately 20 %) set aside for sport hunting and the rules established by the Government of Nunavut, significant economic benefits have facilitated the maintaining of traditional Inuit practices relative to hunting polar bears.

4.4 Hunting as Sharing

The aforementioned aspects defining the hunt can also be found outside the framework of the HTOs, as the meat, fur and even the gifts offered by visitors are distributed amongst the extended family as well as to the community (Dowsley 2009b). Distribution takes place even when private guides take out sport hunters; as they stay with Inuit families and are expected follow local customs. This results in a win-win situation for the community, in the sense that community members were able to go bear hunting, make a fair amount of money and reap the fruits of their labour, not to mention the perpetuation of traditional practices related to hunting (e.g. preparing the hide and making artisanal products). According to Berkes and Armitage (2010), intercommunity sharing has endured as an omnipresent tradition under the banner sport hunting, in spite of the monetary transaction implied in this type of bear hunt, which enables communities to maintain relations internally, as well as with other Inuit communities.

In their ecotouristic adventure, visitors and guides look to hunt large male bears, unlike in subsistence hunting in which drawings and quotas are limited to catching the first bear that comes their way (Dowsley 2009b). They lie in wait for the right bear, although at times hunters come back to community empty-handed. However, after having conducted interviews, Dowsley (2010: 165) stated that the majority nevertheless appreciated their time "in real the wild," and did not regret having gone, as they experienced a "real" hunt learning directly from the Inuit. To lie in wait for the right bear – to be patient – is, to the Inuit, to wait for the bear to accept being hunted and for it to come to them (Dowsley and Wenzel 2008). It is also a good attitude to have towards fellow community members as well as animals. Certain elders, however, criticize this aspect, in that wanting to hunt a bear for commercial gain is disrespectful to the animal. However, the usage of sled dogs, the usual mode of transport, also reaffirms traditions, as they are advantageous in the hunting polar bears (Clark et al. 2008). They find the bear and distract it so the hunters can get into place, which helps facilitate the hunt and minimizes danger of attack. The fact that hunters are not permitted the use of motor vehicles adds an extra layer of respect, as they must hunt at the same pace as the bear.

It is in hunting that both ancestral and current knowledge are reaffirmed, in that they are transmitted and practiced. As previously mentioned, the communities transform products from the hunt and made into objects that often stem from ancestral practical knowledge. These products are usually made by women who also invite their daughters to work with them, to whom the knowledge is then passed down (Dowsley 2009b; Schmidt and Dowsley 2010). Likewise, apprentices learn to become guides. These possibilities, nonetheless, tend to be scarce, as teams look to professionalize by taking on more experienced recruits that can help ensure the longevity of their operation. However, in the HTO-run hunts, student assistants are also employed, where they learn from elders and guides about the bear's behaviour as well as how to understand the environment in general. This is one way in which the culture is revitalized for a new generation, in spite of the greater difficulties they face in accessing this knowledge due to economic obligations and constraints.

As such, sport hunting is currently an important vehicle for the social reaffirmation of all that is related to hunting, as well as a source of financial support so that communities can continue to engage in subsistence hunting. Much like subsistence hunting, sport hunting is steeped in the cosmological conceptions of a reality that correspondingly conditions it. Therefore, the possibility of continuing this practice, in spite of the pressures of the capitalist economy, enables the enactment and revitalization of a significant aspect of the Inuit identity. In this sense, sport hunting does not break with customs, nor is it carried out in a logic of maximization of profit, but rather with respect for the bear and for maintaining community relations.

Significantly, hunting is fertile ground for the sustaining and revitalizing of Inuit identity, as it exists within a larger framework in which multiple forms of relationalities play out. The polar bear hunt is carried out with mutual respect, both amongst Inuit as well as amongst Inuit and bears, based on the foundations of Inuit cosmology. They are based on traditions that have been integrated into practice, and their correspondent application to relationships imbedded with the polar bear mainly reside in hunting, but they also influence an array of dimensions of everyday Inuit life, including but not limited to: exchange, the making of artisanal products and the maintaining of a good attitude. Consequently, sport hunting, in spite of its convergences with the capitalist market, reaffirms and innovates Inuit knowledge, while imbricating it within present-day. It is a tool for Inuit identity.

5 Conclusion

The conflict around the Inuit's hunting of polar bears in Canada underscores several ambiguities regarding the integration of Inuit knowledge into co-management structures, especially considering their omission from governmental scientific reports. The allegations made by different actors in the scientific and political spheres have sparked passion in Inuit communities to defend the hunt based on their own conceptual terms.

The utilization of categories such as Inuit Qaujimajatuqangit has allowed discerning certain aspects of Inuit cosmology that explicate hunting as a means of subsistence and economic development that is intimately connected to Inuit identity. The actor-network theory (Latour 2006) and the notion of relationality (McGrath 2011; Wilson 2008) have reinforced the analysis, insofar as they underscored the prominence of polar bears, in accordance with Inuit cosmology as actors in the hunt.

In this study, we have determined that the dispute amongst scientists and Inuit lies in the primacy of western rationality over Inuit relationality and in the misunderstanding of the latter by the former. Polar bear hunting reflects practices that emanate from a worldview in which the bear is an intelligent and sentient being. It is found in the multi-facetted relationships that the Inuit maintain amongst themselves, as well as with nature, in which the polar bear enjoys a privileged position. Hunting embodies this special relationship; both the Inuit and the polar bear are actors. It is important for the Inuit, as it involves several elements of their identity and makes it possible to reaffirm it in practice. Subsistence hunting is carried out with mutual respect for both the Inuit and the polar bears, and entails an array of savoir-faire directly imbedded in Inuit traditions. Sport hunting, in spite of certain opposition from the communities, is an important aspect of Inuit today's identity as it offers the possibility to raise the money necessary to buy hunting equipment, without which even subsistence hunting would be hard to undertake. However, its practice also allows the enactment of Inuit knowledge and complementary practices.

The debate surrounding the regulation of polar bear hunting in northern Canada clearly cannot be settled without the Inuit point of view being considered, or without Inuit being part of the decision making process. It would be quite contradictory for the international community to ban yet another animal hunted by the Inuit, which helps them reaffirm their identity and ensure food security, when, in those same communities, there are struggles over the extraction of natural resources and for control of the Arctic Ocean passage. These practices have a much greater impact on the habitat of polar bears than does hunting, which is regulated and done in a spirit of mutual respect.

References

Armitage, D., Berkes, F., Dale, A., Kocho-Schellenberg, E., & Patton, E. (2011). Co-management and the co-production of knowledge: Learning to adapt in Canada's Arctic. *Global Environmental Change, 21*, 995–1004.

Berkes, F., & Armitage, D. (2010). Co-management institutions, knowledge, and learning: Adapting to change in the Arctic. *Études/Inuit/Studies, 34*(1), 109–131.

Bell, M. (2002). *Nunavut literacy development in the context of Inuit Qaujimajatuqanginnut (IQ) (Inuit Traditional Knowledge): A discussion paper*. Yellowknife: Inukshuk Management Consultants. http://www.nunavutliteracy.ca/english/resource/reports/paper/cover.htm

Bennet, J., & Rowley, S. (2004). *Uqalurait. An oral history of Nunavut*. Montreal/Kingston: McGill-Queen's University Press.

Clark, D., Tyrrell, M., Dowsley, M., Foote, L. Freeman, M., & Clark, S.G. (2008). Polar Bears, Climate change, and human dignity: Disentangling symbolic politics and seeking integrative conservation policies. International Polar Year 2007–2008, Fall/Winter, 1–6.

Derocher, A. E., Lunn, N. J., & Stirling, I. (2004). Polar Bears in a warming climate. *Integrative and Comparative Biology, 44*(2), 163–176.

Dowsley, M. (2005, September). *Polar Bear as a multiple use resource in Nunavut: Local Governance and common property conflicts*. Paper presented at the Northern Research Forum 3rd Open Meeting: The Resilient North, Yellowknife.

Dowsley, M. (2007). Inuit perspectives on Polar Bears (Ursus maritimus) and climate change in Baffin Bay, Nunavut, Canada. *Research and Practice in Social Sciences, 2*(2), 53–74.

Dowsley, M. (2009a). Community clusters in wildlife and environmental management: Using TEK and community involvement to improve co-management in an era of rapid environmental Change. *Polar Research, 8*, 43–59.

Dowsley, M. (2009b). Inuit-organized Polar Bear sport hunting in Nunavut territory, Canada. *Journal of Ecotourism, 8*(2), 161–175.

Dowsley, M. (2010). The value of a Polar Bear: Evaluating the role of a multiple-use resource in the Nunavut mixed economy. *Arctic Anthropology, 47*(1), 39–56.

Dowsley, M., & Wenzel, G. (2008). The time of the most Polar Bears: A co-management conflict in Nunavut. *Arctic, 61*(2), 177–189.

Freeman, M. M. R., & Wenzel, G. (2006). The nature and significance of Polar Bear conservation hunting in the Canadian Arctic. *Arctic, 59*(1), 21–40.

Gagné, N. (2013). *Being Māori in the city: Indigenous everyday life in Auckland*. Toronto: University of Toronto Press.

Kunuk, Z. (2010). Inuit knowledge and climate change. *Iqaluit, Nunavut: Isuma*. Retrieved September 20, 2012, from http://www.isuma.tv/hi/en/inuit-knowledge-and-climate-change

Latour, B. (2006). *Changer de société – Refaire de la sociologie*. Paris: La Découverte.
Laugrand, F., & Oosten, J. (2002). Perspective inuit et qallunaat: points de vue en interaction. *Études/Inuit/Studies, 26*(1), 9–15.
Laugrand, F., Oosten, J., & Trudel, F. (2000). *Memory and history in Nunavut series*. Nunavut: Nunavut Arctic College.
Martin, T. (2003). *De la banquise au congélateur*. Québec: Presses de l'Université Laval.
Martin, K. (2009). 'Are we also here for that?': Inuit Qaujimajatuqangit – Traditional knowledge. *The Canadian Journal of Native Studies, 29*(1/2), 183–202.
McGrath, J. T. (2011). *Isumaksaqsiurutigijakka: Conversations with Aupilaarjuk towards a theory of Inuktitut knowledge renewal*. Dissertation, Carleton University.
Nadasdy, P. (1999). The politics of TEK: Power and the 'Integration' of knowledge. *Arctic Anthropology, 36*(1–2), 1–18.
Nadasdy, P. (2003). Reevaluating the co-management success story. *Arctic, 56*(4), 367–380.
Nadasdy, P. (2005). The anti-politics of TEK: The institutionalization of co-management discourse and practice. *Anthropologica, 47*(2), 215–232.
O'Neill, T. (2007). *The bear facts; Canada's Inuit say the Polar Bear isn't threatened by global warming or hunting*. Western Standard, 4, 3, April 23 2007. Retrieved May 12, 2014, from http://www.westernstandard.ca/website/article.php?id=2451
Peters, E. (2003). Views of traditional ecological knowledge in co-management bodies in Nunavik, Quebec. *Polar Record, 39*, 49–60.
Randa, V. (1986). *L'ours polaire et les Inuit*. Paris: Selaf.
Rasmussen, K., & Worster, W. (1921). *Eskimo folk-tales*. London/Copenhagen: Gyldendal.
Saladin d'Anglure, B. (1980). Nanuq supermâle, l'ours blanc dans l'espace imaginaire et le temps social des Inuit de l'Arctique canadien. In *L'Ours, l'autre de l'homme*, Numéro spécial, 11 (pp. 63–94). Paris: Centre d'Études mongoles du Laboratoire d'ethnologie et de sociologie comparative.
Schmidt, J. J., & Dowsley, M. (2010). Hunting with Polar Bears: problems with the passive properties of the commons. *Human Ecology, 34*(1), 5–19.
Stuckenberger, A. N. (2010). Les Inuit et le changement climatique [The Inuit and the climate change]. *Études/Inuit/Studies, 34*(1), 377–387.
Taylor, M., Laake, J., Cluff, P. H., Born, E. W., Rosing-Asvid, A., & Messier, F. (2008). Population parameters and harvest risks for Polar Bears (Ursus Maritimus) of Kane Basin, Canada and Greenland. *Polar Biology, 31*(4), 491–499.
Tester, F. J., & Irniq, P. (2008). Inuit Qaujimajatuqangit: Social history, politics and the practice of resistance. *Arctic, 61*(1), 48–61.
Tyrrell, M. (2006). More Bears, less Bears: Inuit and scientific perspectives of Polar Bear populations on the West Coast of Hudson Bay. *Études/Inuit/Studies, 30*(2), 191–208.
Usher, P. J. (2000). Traditional ecological knowledge in environmental assessment and management. *Arctic, 53*(2), 183–193.
Wenzel, G. (1999). Traditional ecological knowledge and Inuit: Reflections on TEK research and ethics. *Arctic, 52*(2), 113–124.
Wenzel, G., & 1. (1989). Sealing at Clyde River, N.W.T.: A discussion of Inuit Economy. *Études/Inuit/Studies, 13*, 3–22.
Wilson, S. (2008). *Research is Ceremony. Indigenous research methods*. Halifax: Fernwood Publishing.
Wong, P. (2010). *Reliability, and tracking techniques of Inuit hunters in estimating Polar Bear characteristics from tracks*. Kingston: Queen's University.

Stéphanie Vaudry is a Phd student in the department of anthropology at the Université Laval. Her phd research focus on how Indigenous groups living in what we now call Uruguay were dispossessed from their lands and how this settler colonial situation influences their current land claims and political strategies. In a previous research, she has explored how Inuit youth and leaders experienced living in the city of Ottawa.

Part IV
Incentives Towards Protected Area Reconciliation and Rights-Based Reform in Governance of Indigenous Territories in the Arctic

Beyond the Protection of the Land, National Parks in the Canadian Arctic: A Way to Actualized and Institutionalized Aboriginal Cultures in the Global

Thibault Martin

Abstract The Canadian government has long excluded Aboriginals from the governance processes of protected areas. However, today Aboriginals peoples, thanks to several judgments of the Supreme Court of Canada have now access to legal tools enabling them to participate equality on park management councils. Despite these legal advances not all co-management models give the same space to Aboriginals people and to their knowledge. As a result management councils express different levels of satisfaction with co-management. Co-jurisdiction is the form of co-management favoured by Aboriginals because it creates legal conditions for an egalitarian partnership based on recognition of their land rights and knowledge that they wish to pass on to future generations.

This chapter examines the management plans of 13 national parks of Canada located in the Arctic. Our study reveals that Aboriginal people feel that culture is the essence of nature and that humans are therefore part of the ecosystem. The protection of nature is therefore part of the duty of men and women. That being said, the environmental protection requires the maintenance of hunting and fishing activities; as such a park must first be a place where Aboriginal culture is living and practiced and not a "pristine natural setting".

Keywords Parks Canada • Governance • Arctic • First Nations • Protected areas

The creation of a national park, especially in regions vulnerable to climate change like the Canadian Arctic, is always a happy event. Beyond protecting biodiversity and the job opportunities they offer, Canada's national parks also contribute to heritage preservation of the land and local cultures. However, until recently they limited access to the land and its resources, especially for aboriginal peoples. This approach to protecting the land led to many protests on the part of the Aboriginal peoples,

T. Martin (✉)
Université du Québec en Outaouais, Gatineau, QC J8X 3X7, Canada
e-mail: thibault.martin@uqo.ca

T.M. Herrmann, T. Martin (eds.), *Indigenous Peoples' Governance of Land and Protected Territories in the Arctic*, DOI 10.1007/978-3-319-25035-9_9

who felt dispossessed of their territorial rights and deprived of the ability to engage in their traditional activities. Beginning in the late 1970s, however, the signing of land agreements and several decisions handed down by Canada's Supreme Court forced public institutions to consider Aboriginal demands, especially in the matter of participation in territorial governance. In the area of natural-resource management, especially the creation of protected areas, co-management appeared to be a way to satisfy the federal government's legal obligations (Samson 2006). In contrast, for the Aboriginals, participation in co-management appeared to be a way to win back power and autonomy (Rodon 2003: 22). Notzke even believes that "There is a possibility that co-management of natural resources will be recognized as a constitutionally entrenched right of Aboriginal people" (Notzke 1995: 187).

This concept of co-management hides a variety of practices ranging from simple consultation to the establishment of true joint management. That is why many studies since the 1990s have focused on the respective merits and limitations of each of the co-management models or have proposed strategies to create forms of co-management that respect Aboriginal rights and needs. However, our purpose is not to assess the limitations or benefits of the various models for co-management of the parks in Aboriginal areas but to approach them from the standpoint of reviving aboriginal governance. In short, we are seeking to understand how participation in managing protected areas can enable Aboriginals to create a form of development that better corresponds to their worldview. As Canada has a very large number and variety of protected-area governance structures, we decided to focus our examination on parks located in Arctic regions, mainly in Inuit territory. There are multiple justifications for this choice. First, the Canadian government is attempting to reach its objective of increasing the total area of protected lands by creating new parks in the most northerly regions, especially because they are less populated and also because the Arctic is especially vulnerable to climate change. The objective is to grant legal protected status to 12 % of Canada's land mass (Canada Parks Agency 2011 (a): Section 2 Leadership and Management). The second reason justifying this choice is that these new co-management structures are often considered innovative and cutting-edge. In fact, it is mainly in the northernmost territories (NWT, Yukon, Nunavut, Nunavik) that the new agreements on co-management of natural resources and environmental protection have given rise to collaboration that gives the greatest importance to Aboriginal partners and knowledge (Samson 2006, Roberts 1996, Campbell 1996, Bailey et al. 1995). According to Nadasdy (2007), these agreements could serve as models for the partnerships to be developed in the south, since several actors have recently developed some respect for Aboriginal knowledge (Nadasdy 2007). We believe it is worthwhile to examine these new forms of co-management.

Political, territorial or economic agreements with the Aboriginal peoples are generally accompanied by emphatic declarations in which governments systematically proclaim that a new "big step forward has just been taken." The researchers who dissect these agreements are sometimes less enthusiastic and often categorize these giant steps as flea-hops forward (Scott and Webber 2001, Mulrennan and Scott 2005), or even backward (Kulchyski 2008). Yet there are very few who take an interest in how the Aboriginals themselves perceive these agreements. The researchers' prevailing opinion is that Aboriginals accept the agreements when they believe they are getting the most that they can from the State. In fact this reading of things is part of the prevailing analytical paradigm in aboriginal studies in Canada, the theory of colonialism/postcolonialism (Martin 2009). It places the study of Aboriginals in a "historical structural perspective" (Frideres 1998: 9), an approach that is heir to theories of conflict. The starting premise in this approach is that the structural framework determining the Aboriginals' condition is the colonial or postcolonial relationship that puts Aboriginals in a defensive situation and ensures that their actions are interpreted as acts of resistance against dispossession/assimilation or territorial/cultural reconquest. For some, the current process of negotiation between the State and the Aboriginals of so-called modern treaties and territorial agreements are merely a sign of the institutionalization of a colonial relationship to a postcolonial one (Neu and Therrien 2003). This analytical grid is far from being uninteresting, as it sheds light on the historical roots and phenomena behind the moral and material conditions of contemporary Aboriginals.

However, this approach tends to see Aboriginal actions as merely a reaction to the action of the State, a State that always has the last word; because of this, it glosses over the "agency" of Aboriginals in silence. However, Aboriginals do not suffer history passively but also act within the colonial relationship and, by their actions, create negotiating spaces that enable them to guide their future according to their own values. In addition, the action they exert on the present is not intended solely, as we have already explained (Martin 2009), to improve the material conditions of their present but also to create the conditions necessary for the advent of a future based on the values of each Aboriginal community, values that originate in the present reading of the cultural heritage. This is therefore a oriented historical action (Martin 2009; Guay 2010).

I came to that conclusion trough the discussions I have had with the Inuit during my visits to the North have led me to realize that their assessments of the agreements and partnerships signed with the State sometimes differ from the researcher's. Indeed, whereas the researcher who assesses the merits and limitations of an agreement judges it according to the "progress" or "gains" – both political and economic – that it brings to Aboriginals, the Inuit instead tend to interpret the same situation from the teleological viewpoint and from a values viewpoint. That is, they assess it not so much in terms of immediate repercussions only but rather as a function of the contribution it will make to the creation of the society they plan to leave to future generations. If this agreement enables them to institutionalize their values and thus project themselves into a future in which they will exercise greater control over themselves, the agreement will be judged positively, and this anticipation will mask its present imperfections. In contrast, an agreement that may, in researchers' eyes, offer real gains for the Aboriginals may very well be rejected by them if they believe it does not allow them to anticipate a future in which they will be completely autonomous but see in it an end point in their autonomy process. The Inuit's vote against the referendum on the *Final agreement on the creation of the Nunavik regional government*, held on 27 April 2011, is an illustration of this. For the entire university community, which includes me (see Chabot 2008), this was an important step for the Inuit in the process of retaking control over their autonomy. On the eve of the referendum, specialists still seemed convinced that the Inuit were going to accept this agreement even though, as Martin Papillion said in an interview with Radio Canada, (Chacon 2011) the Inuit were quite aware that the mere creation of a regional government would not resolve all their problems. Yet the agreement was objected, to the surprise of all observers, by a 66% majority. On the eve of the referendum, in an interview with Radio Canada (*Ibid.*) an Inuk made a comment that reveals some of the reasons that may have pushed the Inuit to vote against the agreement. Indeed, without making any assumptions about the outcome of the referendum, the speaker said nonetheless that many Inuit regretted that there was "no recognition of the Inuit language" in the agreement. In short, it is possible that when they voted, the Inuit weighed the agreement's objective advantages with a subjective anticipation of their future. A future in which they have no institutional guarantee that their values, especially their language, will remain at the core of their society's production. In this chapter, we are going to ask how the new co-management models established in the circumpolar parks, which is to say those in the Territories (Yukon, NWT, Nunavut) and the Torngats Mountains National Park (Newfoundland and Labrador) and Wapusk National Park (Manitoba), may, from the Aboriginal viewpoint, constitute a stage in the formation of their own society, imbedded in the global world.

1 The Place of Aboriginal Peoples in the Governance of Parks in the Canadian Arctic: from Co-management to "Co-jurisdiction"

Before approaching the analysis of the various forms of management present in circumpolar Canada, let us briefly review the concept of co-management and its various forms. The State's withdrawal, which we have been witnessing since the end of the glorious 1930s, means that public institutions are no longer omnipotent or omniscient. This opens up governance spaces, to which new actors are invited, or invite themselves. This encounter/collaboration (desired or sometimes considered a necessary evil) between these various stakeholders and the State takes several forms. We will identify its main characteristics below.

According to Thierry Rodon (2003), co-management is a form of interaction between the State and non-State stakeholders with an interest in protecting an area or in managing a natural resource (Rodon 2003: 115). Co-management refers in fact to a variety of models of interactions ranging from the simple dissemination of information by the State to the sharing of authority (Rodon 2003). The plethora of potential forms of co-management has been described by various authors. In the Aboriginal domain, Berkes et al. (1991) and Berkes (1994) established a typology that takes into account the variety of models of partnership between the State and Aboriginal communities (Table1).

Table 1 Inspired and adapted according to the specific types of park co-management identified by Berkes (1994: 19)

Level of aboriginal participation in park co-management structures		
Co-management framed by land agreements	Partnership	The partners have equal weight; this equality may be enshrined in a treaty. Aboriginal knowledge is recognized and mobilized. Power to administer the park may be delegated to the communities.
	Co-management committee	Communities have an opportunity to participate significantly in the park's governance and operations.
Without agreement	Advisory committee	Structures that enable communities to give an opinion that will be considered.
	Communication	Implementation of institutional mechanisms for exchanges of information between the parties, and for accepting the concerns of Aboriginals.
	Cooperation	Aboriginals are invited on a non-institutional basis to participate in defining the park or in its operations (e.g., use of aboriginal research assistants).
Absence of real co-management	Consultation	Communities are consulted formally, but their opinion is not decisive.
	Information	Communities are informed of the decisions taken upstream.

According to the literature (Andranovich 1995; Berkes 1999; Bornini-Feyerabend 1996; Sneed 1997; Tipa and Welch 2006; Wood and Gray 1991), of all the forms of co-management, it is "collaborative management" or "joint management" that is most likely to create conditions favourable to an equal sharing of power; it is also the type that allows the "joint asset" to be defined not solely by the State but by all stakeholders; it puts aboriginal knowledge and institutional knowledge on an equal footing. This collaborative/joint management echoes the recommendations of the Royal Inquiry Commission on aboriginal peoples, which proposed that the management of aboriginal resources and territories start henceforth from "the principle that the authority of the Aboriginal peoples over their traditional territories is inherent [...]. From this standpoint, agreements should be based on the principle of joint jurisdiction [...which] would lead Aboriginal and non-aboriginal governments to exercise their jurisdiction in a spirit of collaboration, on an equal footing." (Rodon 2003: 122, our translation)

In fact, despite its appeal, co-management, even in its most highly evolved version (joint management) remains first and foremost a means of managing relationships of power among competing stakeholders (Lachapelle 2004). Etymologically the term "partnership" refers to two types of relationships: that of sharing between partners and that of conflict or competition among leading players. This duality is especially noticeable in the area of co-management of natural resources and environmental protection. Indeed, as the State is the defender of the collective heritage, it cannot transfer to anyone, including the Aboriginals, the power to manage fauna and the environment. It "can at the very most delegate to them [Aboriginals] responsibility for applying policies and distributing quotas. Co-management is then a way of integrating Aboriginals into the institutions that manage the State's wildlife" (Rodon 2003: 115, our translation). This maintenance of the State's hegemonic position has pushed some Aboriginals to demand a distinct approach, which they call co-jurisdiction. The first demands in this regard came from speakers before the Royal Inquiry Commission on Aboriginal peoples who doubted the ability of co-management to promote their autonomy, as this testimony from Ernest Ottawa of the Attikamekw Nation indicates: "The question of co-management seems to us inadequate [...]. What we wanted was co-jurisdiction [...]" (cited by Rodon 2003: 122, our translation).

Although, the concept of co-jurisdiction resonates favourably, the fact remains that it does not show up in the literature. Therefore we can offer only a provisional and general definition of it. A relationship of co-jurisdiction might therefore be a situation in which the various parties have powers or rights defined in legislation that ensures that each party is able to exercise exclusive jurisdiction over some aspect of co-management. Such a situation might be the result of land agreements such as, for example, the James Bay and Northern Quebec Agreement. In fact, this agreement gives its beneficiaries exclusive hunting and fishing rights that take precedence over all other legislation, particularly in the area of wildlife protection. The parks that were created at the time in Nunavik are to some extent the result, on the one hand, from a dual jurisdiction of the *Parks Act*, which establishes the overall framework for their creation and administration (in particular, it prohibits hunting and trapping in the parks). And, on the other, the Bay James and Northern Quebec Agreement, which instead guarantees the harvesting rights of the Inuit. The governance of the parks system now being implemented by the Quebec government in Nunavik will therefore have to be achieved in accordance with these two pieces of legislation, one guaranteeing the rights of the Inuit and the other being the instrument put in place by the State to fulfil its mandate in terms of environmental protection.

2 To Ensure the Permanence of Aboriginals' Relationship with Nature

Canada's national parks have a dual mandate: protecting the environment, and developing it for educational and recreational purposes. Reconciling these two objectives has not always been easy, especially when protected areas are established in Aboriginal territory (Héritier 1999; Richard and Pike 1993; Searle 2000; Campbell 1996; Notzke 1994). Indeed, the relationship between the Parks Agency Canada (referred to below as "Parks Canada" or "the Agency") and Aboriginals was at first conflicted, as it excluded them from the protected areas, considering that their way of life interfered with the fulfilment of its mandate to protect (Binnema and Niemi 2006). This prohibition against living or practicing their subsistence activities in "protected" territories was all the more difficult for Aboriginals to accept in that Parks Canada was installing all sorts of infrastructures in order to welcome visitors from around the world. Although things turned out positively for the Aboriginals, the memory of this approach remains etched in their minds. As a result, fear that it will resurface persists, as indicated in this testimony collected at the public hearings held regarding creation of Tursujuq National Park (Nunavik): "Is the only reason for building a park to get visitors to come and play?" (ARK 2008: 105) Tension between Aboriginals and Parks Canada declined as the Agency agreed to recognise that traditional Aboriginal practices and customs did not harm the territory but could even contribute to protecting it.

Of course, as stated above, this new orientation derives in large part from legal obligations defined by the decisions of Canada's Supreme Court requiring Parks Canada to get Aboriginals to participate in the governance of protected areas located in territories over which they hold rights. The first park set up in this context was Ivvavik National park, created in 1984, in the wake of the *Inuvialuit Final Agreement on the Western Arctic*. Vuntut National Park (1995) was set up as part of the *Final Agreement of the Gwitchin Vuntut First Nation;* Auyuittuq National Park has a management structure determined according to the *Nunavut Land Claims Agreement*, to name but these. The signing of these political agreements, which require that Parks Canada bring Aboriginal communities into the governance of protected areas meant that the Agency committed itself to the co-management approach. The change in policy did not happen instantaneously, because, as Berkes and Armitage point out, "It takes time to rework unequal and unjust institutional relationships" (2010: 124).

This transition has however been institutionalized by the adoption of the *Canada National Parks Act (1998)*, and implementation of the *Parks Canada Action Plan in response to the report of the Commission on the Ecological Integrity of the National Parks of Canada* (Canada Parks Agency 2000). The content of the master plans for the various national parks located in the Arctic reflects the evolving nature of the relationship between the Agency and Aboriginal communities. However, this change in attitude has not led to a uniform definition of "co-management". In some cases, there is real equality (at least numerical) between the Aboriginal members and representatives of government institutions who sit on the management board or on the master plan development committee, but in other cases less room is given to Aboriginals. The first scenario is close to a true "partnership" according to Berkes' (1994) definition, whereas in other scenarios co-management is limited to simple "consultation" as in the case of Wood Buffalo Park, where no co-management structure was set up. It is in fact remarkable that the wording used to define this "new relationship" that Parks Canada wishes to develop, or to categorize the management council, varies greatly from one park to another, ranging from simple "Collaboration" to the "Consensus Team" by way of the "Master Plan Development Team" or the "Joint Committee." The same is true for the formation of management councils or master plan development committees, which may vary widely, as in the case of the master plan development committee for Torngat Mountains Park, which consisted entirely of Aboriginals when its first master plan was drafted, whereas the Wood Buffalo committee did not have any (Table 2).

Table 2 Some characteristic features of co-management as it exists in the parks studied

Park name, location and date of latest master plan	Type of co-management (as indicated in official documentation)	Name of the co-management structure	Makeup of the management structure or of the master plan development committee
Torngat Mountains (2010)	Co-management	Management committee	2 members appointed by Makivik
Newfoundland and Labrador			2 members appointed by Nunatsiavut Gvt.
			2 members appointed by Parks Canada
			Chairman appointed jointly by the 3 institutions
Quttinirpaaq (2009)	Participatory management	Joint management committee	3 members appointed by Qikiqtani Inuit Association
Nunavut Territory			3 members appointed by Gvt. of Canada
Auyuittuq (2010)	Participatory management	Joint Inuit-Gvt park planning and management committee	6 reps. of Inuit communities
Nunavut Territory			6 reps of Parks Canada
Sirmilik (–)	No master plan (in 2011)	N. A.	N. A.
Nunavut Territory			
Ukkussiksalik (–)	No information available (in 2011)	N. A.	N. A.
Nunavut Territory			
Ivvavik (2007)	Participatory management	Wildlife management consultative committee (North Slope)	Rep. of an Inuvialuit community, the Yukon Gvt and the Canadian Gvt.
Yukon Territory			
Vuntut (2010)	Co-management	No structure	2 Reps. Gwitchin Vuntut Gvt,
Yukon Territory		Responsibilities shared among institutional partners (Renewable Resources Council of North Yukon, Parks Canada, Government of the Gwitchin Vuntut)	7 reps Yukon Gvt
			3 reps Parks Canada

(continued)

Table 2 (continued)

Park name, location and date of latest master plan	Type of co-management (as indicated in official documentation)	Name of the co-management structure	Makeup of the management structure or of the master plan development committee
Kluane (2010)	Co-management	Management committee	2 reps of Champagne and
Yukon Territory			Aishihik First Nations
			2 Reps. Kluane First Nation
			2 Reps. Gvt of Canada
Tuktut Nogait (2007)	Participatory management	Management committee	2 Reps. Gvt of Canada
Northwest Territories			1 Rep. Inuvialuit Regional Corp.
			1 rep. Game Management Council (GMC)
			1 chairman chosen jointly by minister resp. for Parks Canada, the GMC,
			the Paulatuk Hunters and Trappers Committee and the Paulatuk Community Corp.
Aulavik (2011 preliminary version)	Co-management	Planning and management team	4 reps of Parks Canada
Northwest Territories			2 Reps. Sack Harbor Elders Council
			1 Rep. Sack Harbor Community Corp.
			1 Rep. Sack Harbor Hunter and Trapper Association
Nahanni (2011)	Cooperative management	Nah?a Dehé Consensus Committee	4 Reps. of Dehcho First Nation
Northwest Territories			3 reps of Parks Canada

(continued)

Table 2 (continued)

Park name, location and date of latest master plan	Type of co-management (as indicated in official documentation)	Name of the co-management structure	Makeup of the management structure or of the master plan development committee
Wood Buffalo (2010)	No partnership	Master plan development committee	8 members from Parks Canada
Northwest Territories			
Wapusk (2007)	Collaborative	Management committee	2 Reps. Gvt of Canada
Manitoba			2 Reps. Gvt of Manitoba
			2 Reps. city of Churchill,
			2 Reps. York Factory First Nation
			2 Reps. Fox Lake First Nation

Negotiations and compromises made between the Agency and the Aboriginal communities and with each governance model are the conclusion of "consensus" according to the chosen wording. For example, in the case of Wood Buffalo Park, no partnership with the Aboriginals has been defined. Actually, one of the objectives of the Master Plan is in fact to "define and develop a management structure between Parks Canada and local aboriginal groups and collaborate with Aboriginal groups and with local communities to create a vision statement" (Wood Buffalo Park 2010: 61). In the case of Wapusk Park, the master plan makes no mention of the type of management. It is merely indicated that:

> The Wapusk Park management model stresses collaboration with communities and encourages management and staff to work professionally, transparently, trustingly and respectfully. The Wapusk Management Council influences the park's strategic orientation and guides the process of implementing and observing the agreement on creating the park. (Wapusk Park 2007: section 2. n. p.).

It is in fact interesting to see that the management councils are sufficiently independent in relation to Parks Canada that they are able to reflect, in their formulations, their greater or lesser satisfaction with the co-managementmodel in which they are involved. For example, in the case of Torngats Mountains Park[1] the satisfaction is striking, as the idea that the "Inuit are full-fledged partners" (Torngat

[1] The official name of all parks in Canada is always formed in the same way, with the name of the park, such as "Wapusk," "Tuktut Nogait," etc., followed by "National Park of Canada." To simplify reading, we have abbreviated park names. Hence the "Tuktut Nogait National Park of Canada" becomes "Tuktut Nogait Park."

Mountains Park 2011: no p.) pops up regularly; this equality is even considered an essential factor:

> This park's establishment is the result of collaboration with the Inuit, who are considered full-fledged partners in the project. Parks Canada acknowledges their unique historical and cultural relationship with the land, and pays tribute to it. Their know-how will be integrated into all aspects of the park's management. In fact co-management will play a decisive role here, as we see in it a joint achievement. (Torngat Mountains Park 2011: no p.)

The inclusion on an equal footing of Inuit knowledge (Inuit Qaujimajatuqangit), is a source of great satisfaction as expressed by Willie Nakoolak, chairman of the Wildlife Resources Management Council of Nunavut: "The cooperative, holistic approach adopted by the Inuit and Parks Canada for purposes of managing the park is faithfully reflected in the master plan [...]. This type of partnership will ensure that the Inuit will always be an integral part of the rich, healthy and unique ecosystems that Auyuttuq National Park is home to." (Auyuittuq Park: 2010: iv)

At the contrary, the management council of Aulavik National Park, as part of the updating of its master plan, seeks public input by announcing: "Based on data collected from Inuvialuit co-management institutions, local residents, participants and a broad sampling of the Canadian population, Parks Canada is working to develop Aulavik National Park's master plan (Aulavik Park 2011: no p.). Here it is not a matter of partnership, or of the inclusion of native knowledge on an equal footing, but simply of consultation and data collection based on which Parks Canada is working to develop the master plan." (*Ibid*) In the case of Tuktut Nogait Park, the wording defining the management council's mandate is here again rather unenthusiastic, as it states it was created "so as to counsel the minister of the environment and the other concerned ministers" (Tuktut Nogait Park 2007: iii). As for the park's master plan, it is the result of a compromise reached "despite divergent interests and viewpoints" (*Idem*: 1). The Wapusk Park management council is even more pessimistic; in its introductory message to the master plan, the council states that:

> Overall, the plan represents the consensus reached by the members of the management council. However, as the plan is implemented, the future members of the council and Parks Canada will have to pay special attention to certain matters they have not settled [...] (Wapusk Park 2007: iii).

The transparency of the management councils testifies to their independence from the Agency. In this regard, the Torngat Mountains master plan (2010: 4) stipulates that, "When it offers its opinions, the co-management council does not represent Parks Canada." It is therefore not surprising that the members who sit on these councils express both their satisfaction and their disappointments. We were in fact quite surprised to discover that, beyond the uniform view that Parks Canada takes in its various publications regarding the place Aboriginals should have in co-management, there was in contrast a multitude of views coming out of the management councils.

Our goal in this chapter is not to assess the causes behind the various co-management models instituted by Parks Canada nor even to assess their merits and limitations, but rather to identify what the Aboriginals want to include as their own elements in managing the parks.[2] Since the Aboriginals obtained the place they hold in the co-management structures as the result of negotiations, even "long negotiations," as we can read on the home page of the Vuntut Park web site, we can assume that the elements they want to highlight in the parks are those they consider central in the production of their society. The analysis we did of the master plans enabled us to bring to light three main elements. First, we find a strong determination that the park should contribute to fostering the maintenance of "activities in the territory." It is then a matter of passing on Aboriginal knowledge while updating the culture and placing it in the future. This also involves not isolating the culture, its knowledge and traditions but rather situating it in the contemporary world in relation to non-Aboriginals. For example, the Torngat Mountains Park master plan seeks to make the park a place for:

> [n]ew gatherings in a place outside of time [...]. The objective is to rely on the role of the Torngat Mountains as a traditional gathering place for the Inuit by facilitating contemporary gatherings that celebrate the park as Inuit territory [...] and fostering in people [in this context, non-Aboriginals] a feeling of belonging to the Tongait KakKasuangita SilakKijapvinga through the intermediary of Inuit culture [... The second objective is:] to tell the story of the Inuit, to improve Canadians' understanding of the special bond that binds the Inuit to the rich cultural heritage of the Torngat Mountains. The act of telling the history of the Inuit will promote not just the transfer of Inuit knowledge from the elders to young Inuit, it will also be the main attraction drawing Canadians to discover the power, mystery and adventure of the Torngat Mountains. (Torngat Mountains Park 2010: x)

In most cases, these elements are not presented distinctly but as part of a single project: the actualisation of the culture.

When Aboriginals want the parks to contribute to maintaining activities in the park, they do not simply want to maintain traditional hunting and fishing practices but also to make human presence in the territory current and sustainable, i.e., they want the park to create conditions that update their relationship to the land. Hence Vuntut Park has two "reasons for being," one of which is: "to acknowledge the importance of the Gwitchin Vuntut First Nation's history and culture and to protect the traditional and current use this nation makes of the park" (Vuntut Park 2010: 3). This wording reflects, just as clearly as that on the Torngat Mountains Part cited above, that the relationship to the land that is to be maintained is not defined simply by the traditional use made of it.

Similarly, the vision statement for Quttinirpaaq Park helps us understand why the updating of this relationship to the land is important in Aboriginal eyes. In fact it states that: "[T]he Inuit are an integral part of the Arctic's ecosystems. They manage to live well in these ecosystems, with which they maintain intimate relation-

[2]We proceeded as follows. We first consulted all documents concerning each of the circumpolar parks and identified the various themes that appeared directly concerning Aboriginals: job creation, indigenous knowledge, tourism, land occupancy, elders. We then identified the topics that reappeared in each master plan, and job creation and tourism benefits that appeared only in certain master plans or only marginally were not considered part of the factors that Aboriginals considered essential in the fabric of their society. The analysis consisted of identifying the importance and reasons these topics appear repeatedly.

ships" (Quttinirpaaq Park 2009: 38). This natural belonging to the land is expressed in several other master plans, in particular that for Ivvavik Park, which also affirms that park management relies on "the recognition that humans are part of the ecosystem" (Ivvavik Park 2007: ii). This institutional recognition of Aboriginals' place in the "cycle of nature" is very important, because for them, culture is one with nature (see Martin and Girard 2009). To maintain the relationship with the land is therefore to contribute to updating the culture.

Knowledge and culture transmission is the second element that appears especially important for Aboriginals involved in managing the parks. The master plan for Tuktut Nogait Park states that respect for indigenous knowledge is one of the principles that should guide park management, and it adds that: "[T]he programs to develop the patrimony and maintain the ecological integrity of Tuktut Nogait rely [...] on acquired traditional wisdom and scientific knowledge" (Tuktut Nogait Park 2007: 3). Inuit knowledge (Inuit Qaujimajatuqangit) is also a key aspect of the management of Nunavut's parks, especially in Auyuittuq Park, where "Inuit knowledge is one of the main elements of participatory management" (Auyuittuq Park 2010: 4). It is worth noting that the master plan does not content itself with indicating that Inuit knowledge should be at the core of the management project but devotes several paragraphs to defining it. In this presentation, it states that Inuit knowledge has survived through the centuries "and should be utilized today to ensure a sustainable future" (*Idem*: 3). The Ivvavik Park master plan also states that park management will "respect the culture, language and traditions of the Inuvialuit, both past and present" (Ivvavik Park 2007: 48). The Vuntut Park master plan also refers to language: "Use of the Gwitchin language will be actively promoted." (2010: 30) These commitments attest to the desire of Aboriginals to forge their future from values and knowledge inherited from the past and updated to the present.

One central factor in the corporative plans of Aboriginal communities is the retention of the role of elders as guardians of knowledge and culture. This factor shows up in all the master plans, especially in the vision statement for Tuktut Nogait Park, which stipulates that one of the objectives is to make the park "a tool encouraging elders to pass on their knowledge and culture to young Inuvialuit" (*Idem*: 14). The second objective of the Vuntut Park master plan is entitled "Living and teaching the traditional way of life. This strategy is aimed at preserving the traditional way of life of the Gwitchin Vuntut by helping elders to pass on their knowledge to young people in the community and on the land" (Vuntut Park 2010: viii).

It is worth noting that transmission of culture and knowledge is not simply one benefit of the park, but in many cases one of the reasons, on a par with environmental conservation, that Aboriginals do not merely accept but themselves want to create a park in their territory. Thus the Tuktut Nogait Park master plan states that one "of the reasons [for the park's existence] is to promote a greater understanding and respect for the cultural heritage of the Inuvialuit and the natural environment in which it has evolved" (Tuktut Nogait Park: 3). In short, Aboriginals are in agreement with the creation of a park in their territory when it carries within it the promise of "living and teaching the traditional way of life" (Vuntut Park 2010: vii–vii) so as to "forge the future through the past" (*Idem*). This desire explicitly reveals how

Aboriginals want to orient their society by defining now the future they want to have, a future that is a projection of values rooted in tradition, without being just a retort to it. We have called this exercise in reflexivity a historical action (Martin 2009), a value-oriented action (Guay 2010).

The obligation to utilise indigenous knowledge in managing the park appears in each master plan. The Auyuittuq Park master plan states that:

> Inuit knowledge is one of the main factors in participatory management of Auyuittuq National Park. According to the *Agreement on Repercussions and Advantages for theInuit*, the park management planning program needs to put scientific information and Inuit knowledge on an equal footing […] (Auyuittuq Park 2010: 4).

In the case of this park, the contribution of Inuit knowledge is not a mere objective but a legal obligation, so that:

> the role of Inuit knowledge in research and public-information programs must be examined by the park's Joint Management Committee as part of its approval process. (*Idem*)

It must also be stressed that while Aboriginals want their knowledge to be passed on and used in managing the parks, they nonetheless do not want this knowledge to evolve in a vacuum. The Quttinirpaaq Park strategic plan therefore attaches great importance to the need to create relationships between the Inuit and researchers so as to best exploit the knowledge of each to ensure better management of the park:

> The park's managers will attach as much importance to traditional Inuit knowledge as to scientific information. The Inuit will participate in the park's ongoing research and monitoring programs. (Quttinirpaaq Park 2009: 14)

It is also required that all researchers working in the park communicate the results of their research to the residents of the concerned Inuit communities (Grise Fjord and/or Resolute Bay) (Quttinirpaaq Park 2009: 34); it is also necessary to "ensure that all public information about the park is accessible in Inuktitut" (*Idem*: 40). The place reserved for Inuit knowledge is very important, but it must be connected to scientific knowledge, and it is desired that the population have access to these two sources of knowledge. In this regard, the Vuntut Park master plan is equally explicit:

> Parks Canada, with the collaboration of the Gwitchin Vuntut government, will provide learning opportunities through experiences in the field, such as cultural and scientific camps, for elders, young people, researchers and employees. These opportunities will bring together traditional and scientific knowledge, help the Gwitchin Vuntut pursue their traditional activities on their traditional lands, and promote environmental governance. (Vuntut Park 2011: viii)

In short, it is not so much a matter of "protecting" aboriginal knowledge as of updating it so as to make it serve global knowledge by "combining" it with scientific knowledge.

Furthermore, the objective of perpetuating Inuit culture at the core of the various master plans does not mean that Aboriginals want to step back and protect themselves from other populations but, on the contrary, to profit from the park's creation in order to interact with others. In this regard the authors of the master plan for

Auyuittuq Park remind us that one of the principles of Inuit knowledge (Inuit Qaujimajatuqangit) is "to promote a good state of mind by being open, welcoming and Inclusive" (Auyuittuq Park 2010: 3). James Igloliorte, chairman of the Torngat Mountains Park co-management council, on the occasion of the publication of the park's master plan wrote that it:

> meets the wishes of the Inuit of Nunatsiavut and Nunavik who want to reconnect with the Torngat Mountains, pass on the wisdom of the elders to future Inuit generations, protect this territory forever and share the Inuit history of the Tongait KakKasuangita SilaKijapvinga with Canadian generations to come. (Torngat Mountains Park 2010: v)

The Quttinirpaaq Park master plan stipulates that one of the park's objectives is to ensure that visitors "can enjoy the special relationship that the Inuit have with the land" (Quttinirpaaq Park 2009: 3). For Aboriginals, then, the creation of the parks is a way to encounter others, an encounter that may go so far as to ensure that visitors experience the relationship that binds Aboriginals to the land.[3] For the Inuit, then, it is a matter of opening up their history and culture as well as their privacy to the rest of Canada's population, and to any traveller visiting the Arctic.

3 Placing Culture in the Future

The master plans of Parks in the Canadian Arctic reveal the Aboriginals' plan to modify the parks' strategic orientations so that they contribute to building their society. The purpose of this plan is to protect and develop Aboriginal hunting and fishing practices, folklore, Aboriginal wisdom, historical artefacts and language; it also implies that the Aboriginals' relationship to the land be institutionalized, developed and shared. Aboriginals do not expect park creation to focus on their culture, but they see in the establishment of the parks tools for perpetuating their culture and institutionalizing it in the globalized world. This "exercise" in producing society requires Aboriginals to identify what seems to them to be the core of their way of life and to commit themselves to negotiations with other stakeholders so that these factors can be institutionalized. In short, the action the Aboriginals are now taking is intended to project into the future what they now conceive as the foundations of their culture.

Our analysis also shows that the co-management models that elicit the least enthusiasm among Aboriginals are those in which their need to update their culture is the least developed. Auvalik Park is an example of this, as its master plan (which in any case will be renewed soon) leaves little room for Aboriginals. Indeed, of the eleven objectives included in its vision statement, only two refer to Aboriginals, and these are not exceptional commitments, with one of them merely being the statement of a legal obligation:

[3] That said, nothing in the text explains how a non-Aboriginal can "enjoy this relationship."

> Aulavik will be an exemplary model for subsistence activities in the national parks where traditional use is authorized [...]
>
> scientific research and oral traditions will help better understand the wildlife populations of the park and Banks Isle. (Aulavik National Park of Canada 2002: 10)

The same is true for Wapusk Park which, aside from the words "earth guardian" borrowed from an Aboriginal,[4] leaves little room for their desire to update their wisdom, culture and relationship to the land but instead proposes "visitor-focused management" (Wapusk National Park of Canada 2007: 23). Moreover, this concept of earth guardian is in some way a deviation from its origins; we could even say usurped, as it is not Aboriginals who are deemed guardians of the earth but any visitor to the park:

> In short, our collective and individual responsibility is to ensure that the park's ecological integrity is not disturbed. We are all guardians of the earth. (Wapusk National Park of Canada 2007: 11)

Wapusk, which is one of the first Arctic parks created by Parks Canada using a partnership approach, is a counter-example that helps us understand what it is that Aboriginals do not want to accept as a "co-management" practice. Wapusk Park[5] was officially created in 1996, after 6 years of preliminary consultations. Actual implementation of the co-management was relatively laborious, as the interviews we did among the residents of Churchill and the members of the park's management committee (see Martin 2006; Martin et al. 2004) indicate.[6] The difficult collaboration resulted in the park's management plan not being finalized until 2007, 11 years after it opened. Despite such extended negotiations, the Aboriginals do not seem to have succeeded in placing the updating of their culture at the core of the park project, hence the low level of satisfaction apparent in official texts, especially in the master plan which notes that several areas of disagreement were not settled (Wapusk Park 2007: iii).

The master plans of Canada's circumpolar parks shows that Aboriginals seek recognition of their ancestral rights enshrined in law and try to ensure that parks contribute to updating their culture. Parks Canada has indeed accepted this demand and is now striving to respond favourably to it. However, when the Aboriginals' objectives conflict with other political or economic imperatives (for example, the tourism industry in the case of Wapusk), Aboriginals are not always able to get what they want. Thus there is a certain form of iniquity in the way Aboriginal demands

[4]The expression "earth guardian" is a quote from one of the Aboriginal members of the Management committee, Donald Saunders, York Factory First Nation (Wapusk Park 2007: 6).

[5]This is Canada's 37th national park; it is situated in northern Manitoba in a polar-bear cubbing zone.

[6]This difficult collaboration originates in the fact that the "partners" had to overcome many and varied obstacles. Indeed, unlike certain Arctic parks that are created in areas with undeveloped tourist potential, Wapusk Park was created in a region where tourism is significant. The park is accessible from the city of de Churchill, nicknamed the « Polar Bear World Capital » which receives some 20,000 visitors annually. By way of comparison, Nunavut's three national parks receive fewer than 1,000 visitors per year between the three of them (i.e., an average of 300).

are dealt with, which can be explained by the current legislative context in which ultimate authority over the park is the minister (Weitzner and Manseau 2001; Rodon 2003: 114). This means that the ideal model of "collaborative management" that puts the partners on an equal footing depends on the good will of the minister or of Parks Canada. This comment in the "co-management" section of the Vuntut Park master plan reveals this situation:

> The success of co-management is not the result of the signing of the master plan but rather the result [...] of the determination to aim at the good will and trust of each partner (Vuntut National Park of Canada 1995: 9).

In their study of members of various co-management councils, Weitzner and Manseau find that:

> Many board members stressed that although a strong agreement is in place, what collaborative management means, and how it is operationalized, is largely a question of personality, individuals, and willingness to implement the concept in practice. And [...] to a large extent on the attitude of the park superintendent (Weitzner and Manseau 2001: 255).

In short, to take an idea from Carlsson and Berkes, it could be said that co-management should be considered a learning process more than an established arrangement; the success of co-management therefore depends on the ability of the partners to learn from their mistakes (Berkes 2009). In fact Parks Canada is not committed to defining a specific form of co-management; the "Parks Canada Guiding Principles and Management Policies" stipulate only that:

> In the case of the creation of new parks or reserves, as well as marine conservation areas, or the acquisition of national historic sites, Parks Canada acts within the Canadian political and legal framework relating to the rights of indigenous peoples, as recognized and confirmed by article 35 of the *Constitutional Law of 1982*. Consequently, Parks Canada will consult the concerned communities when a new park is created or a historic site acquired, or in settling a land claim brought by Aboriginals. (Parks Canada 2011: introduction)

This wording in no way implies the creation of an equal partnership. The nature of the partnership may evolve according to the wishes of the stakeholders in attendance, especially those representing the government.

On the other hand, the parks created as part of land agreements are more likely to create the conditions for a genuinely equal partnership, as they give Aboriginals the legal tools to defend their rights. For example, the master plan for Ivvavik Park, created under the *Final Agreement of the Inuvialuit of the Western Arctic*, provides that in the event of a conflict between the *Canada National Parks Act* and the *Act Settling the Land Claim Settlement of the Inuvialuit of the Western Arctic*, the latter takes precedence over the law governing Parks Canada (Ivvavik Park 2007: i). This situation, created by the land agreements, in fact corresponds to what Aboriginals are demanding: co-jurisdiction. The master plan of Torngat Mountains Park also expresses this idea, noting that the various agreements on land claims signed by the Inuit of Labrador and Nunavik define "the limitations and powers" of each (Torngat Mountains Park: 3).

The case of Kluane Park is also very revealing. The desire to protect the region goes back to the 1940s, when a game refuge was created there; then, in 1976, the area was declared a national park reserve. In conservation circles in the 1940s no one was concerned, any more than they were when the national park reserve was created, with Aboriginal rights, so that Aboriginals actually saw their harvesting activities prohibited in the park territory. Still today, despite changes in national legislation, several areas are still off limits to Aboriginal harvesting. These prohibitions were even confirmed by the land agreements signed with the territory's First Nations: *Final Agreement of the Champagne and Aishihik First Nations* (1996) and *Final Agreement of the Kluane First nation* (2003). That said, these agreements changed the balance of power, as they created a legal framework that defines not just the rights of Aboriginals over the park but also a legislative framework allowing good-faith interaction between the parties. This legislative framework led to the creation of a management council that gives the Aboriginals greater control; they no longer have to spend time defending their rights and can work within this framework to steer the park's management so that their rights are better defended. They have therefore undertaken negotiations on these exclusion zones. "For the duration of this master plan, it [the council] will make a recommendation to the Minister of the Environment concerning proposed changes [to the exclusion zones]" (Kluane Park 2010: 11). This is not the situation with the James Bay Cree who opposed the Grande-Baleine hydroelectric project, even though the project was included in the James Bay Agreement. The Cree were successful in their opposition only because the Agreement gave them certainty as to their rights and created opportunities for interaction.

4 Conclusion

Whereas Parks Canada long excluded Aboriginals from the governance processes of protected areas, today Aboriginals, as signatories to political agreements, have legal tools enabling them to access equality on park management councils. Despite these legal advances, however, not all co-management models give the same space to Aboriginals or their knowledge and, as our analysis revealed, the management councils express more or less satisfaction with the management model in which they are involved. Co-jurisdiction is the form of co-management favoured by Aboriginals because it creates legal conditions for an egalitarian partnership based on recognition of their land rights and knowledge. Some years ago, the Institute for the Sustainable Development of the First Nations of Quebec and Labrador introduced the concept of balanced, sustainable development (IDDPNQL 2004). This states the principle that development can be sustainable only if relations between the partners are truly equal, and the *sine qua non* for this is prior recognition of indigenous rights and knowledge (Martin and Girard 2009).

Parks developed under this approach give Aboriginals an opportunity to steer the drafting of master plans so that they can institutionalize the elements of their culture they wish to pass on to future generations. This institutionalization strategy involves cementing the relationship that Aboriginals maintain with the land by updating their knowledge and practices and by affirming that they are part of the ecosystem. However, the Aboriginals who are involved in co-managing parks are not trying to "protect" their culture by shifting it to the background but, on the contrary, by sharing it with other Canadians (and beyond). The same is true for aboriginal knowledge, which they are not so much trying to "protect" as to put to use in service of global knowledge, by promoting partnership with scientists.

In fact, beyond the protection of the land, National Parks in the Canadian Arctic are more for Aboriginals a way to actualized and institutionalized their culture in the global world than a way to "protect" or to freeze traditions.

References

Andranovich, G. (1995). Achieving consensus in public decision making: Applying interest-based problem saving to the challenges of intergovernmental collaboration. *The Journal of Applied Behavioral Science, 31*(4), 429–445.

ARK (Administration régionale Kativik). (2008). *Étude d'impact sur l'environnement et le milieu social*. Projet de parc national des Lacs-Guillaume-Delisle-et-à-l'eau-Claire (Parc national Tursujuq), Kuujjuaq, Administration régionale Kativik.

Bailey, J. L. et al. (1995). Cooperative wildlife management under the Western Inuvialuit Land Claim. In J. A. Bissonette & P. R. Krausman (Eds), *Integrating people and wildlife for a sustainable future. Proceedings of the first International Wildlife Management Congress*. Bethesda: The Wildlife Society.

Berkes, F. (1994). Co-managing: Bridging the two solitudes. *Northern Perspectives, 22*(2-3), 18–20.

Berkes, F. (1999). *Sacred ecology. Traditional ecological knowledge and resource management*. Philadelphia: Taylor & Francis.

Berkes, F. (2009). Evolution of co-management: Role of knowledge generation, bridging organizations and social learning. *Journal of Environmental Management, 90*(5), 1692–1702.

Berkes, F., & Armitage, D. (2010). Co-management institutions, knowledge, and learning: Adapting to change in the Arctic. *Études Inuit, 34*(1), 109–131.

Berkes, F., George, P., & Preston, R. J. (1991). Co-management: The evolution in theory and practice of the joint administration of living resources. *Alternatives, 18*(2), 12–19.

Binnema, T., & Niemi, M. (2006). 'Let the line be drawn now' wilderness, conservation and the exclusion of aboriginal people from Banff National Park in Canada. *Environmental History, 11*, 724–750.

Bornini-Feyerabend, G. (1996). *Collaborative management of protected areas: Tailoring the approach to the context*. Gland: IUCN.

Campbell, T. (1996). Co-management of aboriginal resources. *Information North, 22*(1). Retrieved April, 2011, from http://arcticcircle.uconn.edu/NatResources/comanagement.html

Canada Parks. (2000). *Unimpaired for future generations?* (A call to action conserving ecological integrity with Canada's national parks, Vol. I). Ottawa: Canada Parks.

Canada Parks. (2011). *Parks Canada guiding principles and operational policies*. Ottawa. Canada Parks Retrieved April, 2011, from http://www.pc.gc.ca/eng/docs/pc/poli/princip/index.aspx

Chabot, M. (2008). *Regards croisés sur le projet de gouvernance du Nord québécois: Thibault Martin et Donat Savoie*, Développement social, Montréal, Institut national de santé publique du Québec, 9 (1), 29–33.

Chacon, G. (2011). Interview by Geneviève Chacon, broadcasted on Radio Canada on April 27, 2011 (17: 00).

Frideres, J. S. (1998). *Aboriginal Peoples in Canada: Contemporary Conflicts* (5th ed.). Scarborough: Prince Hall/Allyn and Bacon.

Guay, C. (2010). *La rencontre des savoirs à Uashat Mak Mani Utenam: regard sur la pratique des intervenants sociaux innus*. Thèse de doctorat en sciences sociales appliquées, Gatineau, Université du Québec en Outaouais.

Héritier, S. (1999). De la conquête territoriale à la conquête environnementale : quelques processus d'appropriation du territoire canadien (à travers l'exemple des parcs de Banff et Jasper). *Études canadiennes* (Canadian Studies), *47*, 73–94.

Institut du Développement Durable des Premières Nations du Québec et du Labrador. (2004). *Mémoire déposé à la Commission d'étude scientifique, publique et indépendante chargée d'examiner la gestion des forêts du domaine de l'état*. Wendake, Assemblée des Premières Nations du Québec et du Labrador.

Kulchyski, P. (2008). A step back: The Nisichawayasihk Cree Nation and the Wuskwatim project. In T. Martin & S. Hoffman (Eds.), *Power struggles: Hydro-development in Quebec and Manitoba* (pp. 129–144). Winnipeg: University of Manitoba Press.

Lachapelle, R. (2004). Le développement des communautés : enjeux et opportunités. Communication à la journée d'étude intersectorielle de la Haute-Yamaska, Granby, January 15.

Martin, T. (2006). Co-management of a national park: The Wapusk National Park's experience. *Études canadiennes* [Canadian Studies], *61*(2), 139–148.

Martin, T. (2009). Pour une sociologie de l'autochtonisme. In N. Gagné, T. Martin, & M. Salün (Eds.), *Autochtonies. Vues de France et du Québec* (pp. 431–454). Québec: Les Presses de l'Université Laval.

Martin, T., & Girard, A. (2009). La forêt matrice de culture. *Recherches amérindiennes au Québec, 39*(1-2), 61–70.

Martin, T., Falvo, L., & Chotka, M. (2004). *Environment and economic development: Co-managing a National Park while stimulating Community Development in Churchill (MB)*. Winnipeg: Department of Sociology, University of Winnipeg.

Mulrennan, M. E., & Scott, C. H. (2005). Co-management – An Attainable Partnership? Two Cases from James Bay, Northern Quebec and Torres Strait, Northern Queensland. *Anthropologica, 47*, 197–214.

Nadasdy, P. (2007). The gift in the animal: The ontology of hunting and human–animal sociality. *American Ethnologist, 34*(1), 25–43.

Neu, D., & Therrien, R. (2003). *Accounting for genocide: Canada's bureaucratic assault on aboriginal people*. Black Point: Fernwood Publishing.

Notzke, C. (1995). A new perspective in aboriginal natural resource management: Co-management. *Geoforum, 26*(2), 187–209.

Park Canada Aulavik. (2002). *Park management*. Ottawa: Canada Parks.

Park Canada Aulavik. (2011). *Web site of the National Park of Aulavik*. Retrieved April, 2011, from http://www.pc.gc.ca/fra/pn-np/nt/aulavik/plan.aspx

Park Canada Auyuittuq. (2010). *Park management*. Ottawa: Canada Parks.

Park Canada Ivvavik. (2007). *Park management*. Ottawa: Canada Parks.

Park Canada Kluane and Reserve. (2010). *Park management*. Ottawa: Canada Parks.

Park Canada Nahanni. (2011). Web site of Nahanni National Park Reserve of Canada. Retrieved October, 2015 from http://www.pc.gc.ca/eng/pn-np/nt/nahanni/index.aspx

Park Canada Quttinirpaaq. (2009). *Park management*. Ottawa: Canada Parks.

Park Canada Torngat Mountains. (2011). *Park management*. Ottawa: Canada Parks.

Park Canada Torngat Mountains (Tongait KakKasuangita SilakKijapvinga). (2010). *Park management*. Ottawa: Canada Parks.

Park Canada Tuktut Nogait. (2007). *Park management*. Ottawa: Canada Parks.

Park Canada Vuntut. (2010). *Park management*. Ottawa: Canada Parks.

Park Canada Vuntut. (2011). *Web site of the National Park of Vuntut*. Retrieved April, 2011, from http://www.pc.gc.ca/fra/pn-np/yt/vuntut/index.aspx

Park Canada Wapusk. (2007). *Park management*. Ottawa: Canada Parks.

Park Canada Wood Buffalo. (2010). *Park management*. Ottawa: Canada Parks.

Richard, P. R. & Pike, D.G. (1993). Small Whale co-management in the Eastern Arctic: A case history and analysis. *Arctic, 46*(2). Retrieved April, 2011 from http://www.highnorth.no/library/Policies/National/la-cl-ag.htm.

Roberts, K. (Ed.), (1996). Circumpolar aboriginal people and co-management practice, Inuvik, Northwest Territories. In Workshop Proceedings (November 20–24, 1995). Arctic Institute of North America. Calgary. The University of Calgary and Joint Secretariat-Inuvialuit Renewable Resource Committee.

Rodon, T. (2003). *En partenariat avec l'État. Les expériences de cogestion des Autochtones du Canada*. Québec: Les Presses de l'Université Laval.

Samson, R. (2006). Nature et Culture dans les parcs nationaux: Reconnaître les liens. *Revue d'histoire de l'Amérique française, 60*(1–2), 191–203.

Scott, C. H., & Webber, J. (2001). Conflicts between Cree hunting and sport hunting: co-management decision making at James Bay. In H. Scott Colin (Ed.), *Aboriginal autonomy and development in northern Quebec and Labrador*. Vancouver: UBC Press.

Searle, R. (2000). *Phantom Parks: The struggle to save Canada's national parks*. Toronto: Key Porter Books.

Sneed, P. (1997). Towards co-management of national parks in Canada and the U.S.A. In S. Stevens (Ed.), *Conservation through cultural survival: Indigenous peoples and protected areas* (pp. 135–154). Washington, DC: Island Press.

Tipa, G., & Welch, R. (2006). Comanagement of natural resources: Issues of definitions from an indigenous community perspectives. *The Journal of Applied Behavioral Science, 42*(3), 373–391.

Weitzner, V., & Manseau, M. (2001). Taking the pulse of collaborative management in Canada's national park and national park reserves: Voice from the field. In D. Harmon (Ed.), *Crossing boundaries in park management: Proceedings of the 11th conference on research and resource management in parks and on public lands*. Hancock: The George Wright Society.

Wood, D., & Gray, B. (1991). Toward a comprehensive theory of collaboration. *The Journal of Applied Behavioral Science, 27*(2), 139–162.

Recognition of Indigenous Lands Through the Norwegian 2005 Finnmark Act: An Important Example for Other Countries with Indigenous People?

Øyvind Ravna

Abstract International law includes treaties and declarations that commit the national states to protect the culture and livelihood of indigenous peoples. Of particular interests is The International Labour Organization (ILO) Convention No. 169 concerning indigenous and tribal peoples in independent countries and the United Nations Declaration on the Rights of Indigenous Peoples (UNDRIP). Norway is one of 22 countries that is party to the ILO 169. In this chapter, the commitment to identify and recognize indigenous people's lands and natural resources in relation to the indigenous Sámi in the Nordic Countries will be examined. This commitment applies in particular to Norway, which is the only country with a Sámi population who has ratified the ILO Convention. The commitments imposed to Norway thus raises several key issues regarding identification of indigenous people's lands, including to what extent the Sámi laws and customs have significance as legal sources in such processes, and how to the state must involve the indigenous party in the process.

Keywords Finnmark Act • Sámi • ILO Convention No. 169 • Indigenous rights

1 Introduction

International law includes treaties and declarations that commit the national states to protect the culture and livelihood of indigenous peoples. Of particular interests is The International Labour Organization (ILO) Convention No. 169 *concerning*

Ø. Ravna (✉)
Sámi and Indigenous Law, Faculty of Law, University of Tromsø –
The Arctic University of Norway, Tromsø, Norway
e-mail: oyvind.ravna@uit.no

T.M. Herrmann, T. Martin (eds.), *Indigenous Peoples' Governance of Land and Protected Territories in the Arctic*, DOI 10.1007/978-3-319-25035-9_10

indigenous andtribalpeoples in independent countries[1] and the United Nations *Declaration on the Rights of Indigenous Peoples* (UNDRIP).[2]

In this article, the commitment to identify and recognize indigenous people's lands and natural resources in relation to the indigenous Sámi in the Nordic Countries will be examined.[3] This commitment, which are ensured under the ILO Convention no. 169, applies in particular to Norway, which is the only country with a Sámi population who has ratified the ILO Convention. The commitments imposed to Norway raises several key issues regarding identification of indigenous people's lands, including to what extent the Sámi customary law have significance as legal sources in such processes, and how to the state must involve the indigenous party in the process.

To follow up the commitments in the ILO Convention, the Norwegian Parliament in 2005 adopted an act aiming to contribute to the identification process. The act is limited to frame the County of Finnmark, which is the most central part of the Sámi traditional lands (Sápmi) in Norway; therefore the name *the Finnmark Act* (Fm Act).[4] The Act is emphasized as an example for the other Nordic Countries by the UN Special Rapporteur on the rights of indigenous people, *James Anaya.* In his report on the situation of the Sámi people in the Sápmi region of Norway, Sweden and Finland, he has pronounced that the Finnmark Act is "an important protection for the advancement of Sámi rights to self-determination and control over natural resources at the local level, setting an important example for the other Nordic countries".[5] However, the Act and the adopting process has been controversial;

[1]Adopted June 27, 1989 and coming into force September 5, 1991. Retrieved May 10, 2012, from http://www.ilo.org/ilolex/cgi-lex/convde.pl?C169. The Convention is presently ratified by 22 states, among them Norway, see *ibid.*

[2]See http://www.un.org/esa/socdev/unpfii/documents/DRIPS_en.pdf Retrieved May 12, 2012. The Declaration was adopted by the United Nations General Assembly, 107th plenary meeting, September 13, 2007, which voted on the adoption. The vote was 143 countries in favor, 4 against, and 11 abstaining. The four member states that voted against, Australia, Canada, New Zealand and the United States, have later endorse the declaration. Among the abstaining countries is The Russian Federation. Retrieved ay, 12, 2012 from http://social.un.org/index/IndigenousPeoples/DeclarationontheRightsofIndigenousPeoples.aspx

[3]The Sámi live in the northern and central parts of Norway, Sweden and Finland, and the Kola Peninsula in Russia, where they in the three first mentioned countries, is the only indigenous people. They consist of 50,000–80,000 peoples earning their livelihood from both marine and terrestrial industries such as reindeer husbandry, agriculture and coastal fishing, see Harald Gaski in Store Norske Leksikon, http://snl.no/samer. For more information about the Sámi, see the Sámi Parliament's web page. Retrieved May 4, 2012, from http://www.samediggi.no/artikkel.aspx?AId=3688&MId1=3487. The Article 14 of the ILO Convention no. 169 is of particular interests, which imposes the contracting States to identify and recognize indigenous peoples' traditional lands.

[4]Lov (Act) 17. juni 2005 nr 85 om rettsforhold og forvaltning av grunn og naturressurser i Finnmark fylke (Finnmarksloven) [Act 17 June 2005 No. 85 relating to legal relations and management of land and natural resources in the county of Finnmark]. English translation. Retrieved April 20, 2012, from http://www.ub.uio.no/ujur/ulovdata/lov-20050617-085-eng.pdf

[5]James Anaya, *The situation of the Sami people in the Sápmi region of Norway, Sweden and Finland,* Report on the situation of human rights and fundamental freedoms of indigenous

conservative politicians have criticized it for giving the Sámi too big influence, while the Sámi Parliament and Sámi NGOs have taken the opposite standing.[6]

The aim of this article is not to query any of those opinions, but to analyze the legal development that forms the bases for the opinions, i.e. the legislature process of the Finnmark Act. In addition, the analysis also frames the outcome of the process, including how Norway fulfills the commitments in ILO Convention no. 169 Article 14 and other legislation imposed to identify and recognize the lands the Sámi are presumed to own and possess.

Sources for the analysis are mainly legislation, including preparatory works, supported by case law and legal literature. The theme is actualized since it proposes similar schemes for clarification in the Sámi areas south of Finnmark,[7] and since the Finnmark Commission recently has delivered its first report.[8]

The Finnmark Act is a land code consering legal relations and management of land and natural resources in the county of Finnmark. Section One of the *Act* outlines: that "The purpose of the *Act* is to facilitate the management of land and natural resources in the County of Finnmark in a balanced and ecologically sustainable manner for the benefit of the residents of the county and particularly as a basis for Sámi culture, reindeer husbandry use of non-cultivated areas, commercial activity and social life".

The Finnmark Act is thus more than a land code, it is also a law aimed to protect indigenous lands and culture, born through controversies and consolations with the Sámi.

With its entry into force on 1 July 2006, the Act transferred all "unsold state lands", which represent approximately 95 % of the county's total area and almost all the outlying and mountainous areas, to an ownership body called the *Finnmark Estate,*[9] cf. Fm Act S. 6. This body is partly ruled by the Sámi Parliament and partly by the Finnmark County Counsel, which both has three members of totally six of the board, cf. Fm Act S. 7, Para. 2.

---

people (2011), Para. 44 Retrieved May2,2012 from http://unsr.jamesanaya.org/docs/countries/2011_report_sami_advance_version_en.pdf

[6] The different opinions was clearly shown in several Northern Norwegian Newspaper during the process of adopting the act, see Øyvind, R. (2004). Forslaget til 'Finnmarkslov' og bygdefolks rettigheter. Kritisk juss, 1 (30), 35–57.

[7] See NOU 2007: 13 *Den nye sameretten,* pp. 31–68.

[8] See Finnmarkskommisjonen, *Rapport felt 1 Stjernøya / Seiland,* March 20, 2012. As the report was submitted at the time this paper was to be completed, the findings in the report is no topic here. For a brief analysis of the findings, see: Øyvind Ravna, "The First Investigation Report of the Norwegian Finnmark Commission", *International Journal on Minority and Group Rights,* 3 2013 (20), pp. 443–457. The Finnmark Commission has afterwards delivered three more reports; *Rapport felt 2 Nesseby,* February 13, 2013 and *Rapport felt 3 Sørøya,* October 16, 2013 and *Rapport felt 5 Varangerhalvøya Øst*, June 24, 2014.

[9] In the original law text the name is *Finnmarkseiendommen – Finnmárku opmodat* (the last in Sámi language). In Fm Act S. 6 the Finnmark Estate is defined as "a separate legal entity with its seat in Finnmark, which shall manage the land and natural resources... as the owner in accordance with the purpose and provisions of the Act in general".

Of specific interest for this presentation, The Finnmark Act initiates a process of legal identification and recognition of land rights for areas that previously were considered to be state-owned land,[10] aiming to identify and determine ownership and usage rights based in immemorial usage etc., both of individual and collective characters. The investigation is to be performed by a body called The Finnmark Commission, cf. Fm Act Section S. 29 Para. 1, which recently has completed its first investigation, while a special court, the Land Tribunal for Finnmark,[11] is to settle disputes arising from the investigation of the Commission, cf. Fm Act S. 36 Para 1.

2 The Preparatory Work of the *Finnmark Act*

2.1 A Backdrop

Unlike in most of North America, the Indigenous lands in the Nordic Countries have never been subject for treaties between the European colonists and Indigenous people. This can be explained by the fact that the Sámi and the Norse have lived side by side for almost 1000 years, and that the Sámi not have had a tradition to defend their land with arms. It might also be explained in the system of the old feudal Europe, where the king regarded his power to be supreme, ruling the country as his private property with a far-reaching right to subjugate serfdom and collect taxes. Instead of treaties, the Sámi base their rights on immemorial usage of lands (historically use) and more recently also in international human rights law.

As a result of the Sámi struggle for recognition, Norway step by step has undertaken obligations to protect the Sámi language, culture and way of life. The Constitutional amendment of 1988,[12] the ratification of ILO Convention No. 169 concerning indigenous and tribal peoples in independent countries of 1989 (ratified by Norway in 1990), the 1999 Human Rights Act,[13] and the promotion of the UN indigenous declaration on indigenous peoples' rights (2007) are all parts of that picture.

[10]An overview of the *Finnmark Act* and the procedural law requirements can be found in Øyvind Ravna, 'The Process of Identifying Land Rights in parts of Northern Norway: Does the Finnmark Act Prescribe an Adequate Procedure within the National Law?" *Yearbook of Polar Law,* Brills (3) 2011 pp. 423–453 on pp. 425–429.

[11]The term "Uncultivated Land Tribunal for Finnmark" is often used in English translation. It does not reflect the Sámi point of view, since livelihood and cultural activities historically have not depended on actual land cultivation. The outlying land and mountainous areas are consequently Sámi cultural land. Therefore the more neutral form, the Land Tribunal for Finnmark, is used.

[12]Kongeriget Norges Grundlov 17. mai 1814 [The Constitution of the Kingdom of Norway 17 May 1814], Article 110 a. English translation. Retrieved May 3, 2012, from http://www.stortinget.no/en/In-English/About-the-Storting/The-Constitution/The-Constitution

[13]Lov 21. mai 1999 nr. 30 om styrking av menneskerettighetenes stilling i norsk rett (menneskerettsloven) [Act 21 May 199 No. 30 relating to the strengthening of the status of human rights in Norwegian law], In English translation. Retrieved May 3, 2012 from http://www.ub.uio.no/ujur/ulovdata/lov-19990521-030-eng.pdf

The Finnmark Act and the identification process come out of this development, too, and can as such be seen as a response to many years of struggle by the Sámi, and due in part to the infamous conflict surrounding the construction of *the Alta-Kautokeino hydro power plant* in the 1970s, which included a proposal to flood the Sámi village of Maze.[14] But the Sámi cultural and legal awareness was not only stimulated by the plans of building a power plant in the heart of *Sápmi* in the 1970s, but also out of international contacts through the *International Work Group for Indigenous Affairs (IWGIA)* and movement among more numerous indigenous peoples, as the American First Nations and the *American Indian Movement (AIM)*.[15] The Sámi got support in theoretical analysis questioning the State ownership of the Sámi lands, too.[16] Together, this culminated in the Alta case at the end of the decade, which prompted the government to establish the *Sámi Rights Committee* in 1980 to investigate the Sámi legal status.

2.2 The Draft of the Sámi Rights Committee and the Governmental Response

The investigatory work that took place under the umbrellas of the Sámi Rights Committee, which was a law committee, was the first formal step in forming the Finnmark Act.[17] The investigation led to acknowledgement that state ownership of

[14] Ot Prp [Proposition to the Parliament] Nr 53 (2002–2003) *Om lov om rettsforhold og forvaltning av grunn og naturressurser i Finnmark fylke* (Government bill for the Finnmark Act). See also James Anaya, supra note 5, Para 18. For more reading about the Alta Case, see Galdu, *The damming of the Alta-Kautokeino Watercourse (The Alta Case).* Retrieved May 3, 2012, from http://www.galdu.org/govat/doc/eng_damning.pdf and Svein S. Andersen & Atle Midttun (1985). Conflict and Local Mobilization: The Alta Hydropower Project, Acta Sociologica, 28, 317– 335.

[15] See Henry Minde, "Challenge of indigenism: the struggle for sami rights and self-government in Norway in 1960–1990", Svein Jentoft, Henry Minde and Ragnar Nilsen (eds.), *Indigenous Peoples: Resource Management and Global Rights,* (Ebourn Academy Publishers, Delft 2003), pp. 75–104 at p. 81 and Henry Minde, *The International Movement of Indigenous Peoples: an Historical Perspective,* Center for Sámi Studies, University of Tromsø. Retrieved May 3, 2012 from http://www.sami.uit.no/girji/n02/en/003minde.html

[16] Sverre Tønnesen, (1972) *Retten til jorden i Finnmark. Rettsreglene om den såkalte "Statens umatrikulerte grunn" – en undersøkelse med særlig sikte på samenes rettigheter*, Universitetsforlaget, Bergen.

[17] The Sámi Rights Committee had a mandate of four points where the first was to examine "the question about the Sámi people's legal position in relation to land and water" including a consideration of the need for changes in current law, in where it could submit proposals for new regulations including new legislation. The next two points where to examine and suggest "how to secure the Sámi population opportunities to utilize natural resources in their areas of habitation, while also recognizing the non-Sámi population's interests" and to examine the need of a constitutional protection of Sámi culture and language. The last point was of administrative and economical character, see NOU [Norwegian Public Report] 1984: 18 *Om samenes rettsstilling*, pp. 43–44. The particular proposal for a Finnmark Act was worked out in NOU 1997: 4 N*aturgrunnlaget for samisk kultur.*

unsold land in Finnmark was based upon a legal opinion which the Norwegian State no longer fully could support. Although a subcommittee of legal experts under the Sámi Rights Committee, in 1993 concluded that the Norwegian state was the owner of the unsold land in Finnmark, both the coastal parts and in the interior (Sámi) parts of the County, it raised a fundamental question regarding the legitimacy of that ownership, stating that *it might be based on a misunderstanding that is difficult to excuse*.[18]

Even though the Sámi Rights Committee undertook a general discussion of the legal basis for natural resources in Finnmark, it did not assess actual ownership and rights of use acquired by the Sámi and others, but based their position on the finding of the subcommittee.

However, the Sámi Rights Committee found that the Sámi and other locals had certain rights of use and proposed an act for the management of the land in Finnmark, whereby the title should be transferred from the State Forest Company (Statskog SF) to an independent ownership body called the *Finnmark EstateManagement* (Finnmark grunnforvaltning). This body should be controlled by a board appointed in part by Finnmark County Council and in part by the Sámi Parliament.

The Sámi Rights Committee also proposed a governance model in which the locals would be given influence over management of renewable natural resources through locally-appointed "outfields boards". It further proposed that so called *community commons* (bygdebruksområder) should be identified and recognized, based on local traditional usage, which could be considered as a kind of "modern *siida* system."[19]

In addition, the Committee also made a proposal for a procedure to identify such commons. It did not propose a commission or tribunal as now prescribed in the Finnmark Act,[20] but suggested instead the community commons to be determined by a local committee appointed for each municipality. This was reasoned in that such identification demanded local knowledge.[21]

The question of Sámi rights to natural resources in Finnmark was controversial. Six years after the Sámi Rights Committee submitted its draft, the Bondevik government presented a bill for a Finnmark Act. Based on the findings of the subcommittee of the Sámi Rights Committee, the Government agreed upon that the State ownership could not be upheld in full.[22] It also accepted that the lands of

[18] NOU 1993: 34 *Rett til og forvaltning av land og vann i Finnmark*, p. 263 (My italics).

[19] NOU 1997: 4, p. 241. In former times *siida* was a Sámi community which managed a physically-determined territory; see Erik Solem, *Lappiske rettsstudier*, pp. 81–84. Today the concept is used for a family-related working unit in reindeer husbandry, c.f. Reindeer Husbandry Act, S. 51–56.

[20] This is worth noting, as the Uncultivated Land Commission for Nordland and Troms (1985–2004), which was a tribunal mandated to determine boundaries between State and Private lands in the Counties of Nordland and Troms, at that time (1990s) was at the peak of its productivity.

[21] The Sámi Rights Committee also proposed the land consolidation court would be the appeal body, since the procedure to determine boundaries under Section 88 and 89 of the 1979 Land Consolidation Act seemed the most natural process form when it came to delineation questions.

[22] Ot.prp. nr. 53 (2002–2003) *Om lov om rettsforhold og forvaltning av grunn og naturressurser i Finnmark fylke* (The *Finnmark Act*), p. 43. An important reason for the governmental acknowledge-

Finnmark could be subject to "private or collective rights based on prescription or immemorial usage".[23]

This implied that the government followed up the proposal to transfer the ownership to a body controlled by the Sámi Parliament and the Finnmark County Council. But the government did not accept the Sámi Rights Committee proposal for local community commons management of the outlying fields. This was justified in that the outfield resources should be managed uniformly, not to "harm a desired and appropriate allocation of resources in their entirety."[24]

The proposal of the Sámi Rights Committee to identify community commons was an attempt to recognize the rights to "lands and waters" in the Sámi areas.[25] Although the government in section 5 of the draft act acknowledged that Sámi and others had acquired rights by prescription and immemorial usage on the former state land,[26] the proposal to identify lands and rights was omitted from the draft. No other suggestions were either made to identify such rights. The Government aimed instead "to make good arrangements for the rights and the management of lands and waters in Finnmark by law rather than by dispute resolution in the courts."[27] Transferring the land to the *Finnmark Estate* was the way to reach that aim and to follow up Norway's commitment in the ILO Convention no. 169. Accordingly it did not put forward any proposals or procedures to conduct an identification and recognition process.

The draft act was met with considerable criticism from the Sámi Parliament, who argued that the bill was not in accordance with obligations under international law, especially the ILO Convention No. 169 to identify indigenous people's traditional lands.[28] As a result the Norwegian Parliament by the Standing Committee of Justice asked for an independent assessment of the draft act, which the Professors Geir Ulfstein and Hans Petter Graver were engaged to undertake. They concluded that the government's proposals on key points were insufficient to meet ILO Convention no. 169. In relation to Article 14, they found that if the *Finnmark Act* shall meet ILO Convention requirements for recognition of land rights, "the decision rules must be changed in such way that the Sámi are secured the control according to an ownership position. If this not relevant for the entire county, the particular Sámi areas need to be identified with a view to ensuring the Sámi the control and rights to these areas".[29]

---

ment was the Norwegian ratification of the United Nations' International Labor Organization (ILO) Convention No. 169 concerning Indigenous and Tribal Peoples in Independent Countries in 1990.

[23] Cf. the Proposal for a Finnmark Act Section 5(1), see Ot.prp.nr 53 (2002–2003) p. 122. All translations of quotations, except for the one of Finnmark Act, are done by the author.

[24] Ot.prp. nr. 53 (2002–2003), pp. 98–99.

[25] See also Jon Gauslaa, "Lov om rettsforhold og forvaltning av grunn og naturressurser i Finnmark fylke (finnmarksloven)" Gyldendal rettsdata, note 3.

[26] Ot.prp. nr. 53 (2002–2003), *Ibid.*, p. 8 and 122, cf. the draft S. 5.

[27] *Ibid.*, p. 97.

[28] Innst. O. nr. 80 (2004–2005), p. 14.

[29] Geir Ulfstein and Hans Petter Graver, *Folkerettslig vurdering av forslaget til ny finnmarkslov.* Retrieved April 28, 2012 from http://www.regjeringen.no/nb/dep/jd/dok/rapporter_planer/rapporter/2004/folkerettslig-vurdering-av-forslaget-til.html?id=278377

The criticism, in particular the requirement for the bill to comply with international indigenous people law, initiated a new era for constitutional practice in Norway.[30]

2.3 The Final Preparation of the Act; Consultations between the Parliamentary Standing Committee of Justice and the Sámi Parliament

On the initiative of the Parliamentary Standing Committee of Justice, four consultations with the Sámi Parliament and Finnmark County Council took place in 2004 and 2005. The majority of the Standing Committee of Justice pointed out that "Norway's international obligations to consult the Sámi are thus included in Parliament's work. This is a constitutional innovation."[31]

The consultations led to rather extensive changes in the draft, which included a new first paragraph of Section 5 stating "the Sámi have collectively and individually through prolonged use of land and water acquired rights to land in Finnmark." That statement is said to represent a principle and political recognition that such rights exist.[32]

Due to these consultations the majority of the Standing Committee, with the exception of the members from the Progress Party and the Socialist Left Party, acknowledged that identification of existing rights must be included as a key element in the *Finnmark Act*[33] proposing established "a surveying commission and a judging tribunal to identify existing rights to land and water in Finnmark."[34] The identification and recognition procedure is regulated in the Finnmark Act chapter 5 (SS. 29–43).

The commission is mandated to investigate and clarify the legal situation on the lands the Finnmark Estate has taken over from the State Forest Company. The report of the Commission will, according to the majority of the Standing Committee of Justice, provide a good basis for people in Finnmark to make up their mind whether conflicts over land rights actually exist.[35] The intention is that the ambiguities and

---

[30] In 2005 there was also entered into a consulting agreement between the Norwegian Government and the Sámi Parliament to contribute to a practical implementation of the state's international legal obligation to consult the Sámi, see Prosedyrer for konsultasjoner mellom statlige myndigheter og Sametinget. Retrieved May 12, 2012, from http://www.regjeringen.no/nb/dep/fad/tema/samepolitikk/midtspalte/prosedyrer-for-konsultasjoner-mellom-sta.html?id=450743

[31] Innst. O. nr. 80 (2004–2005), p. 15.

[32] Innst. O. nr. 80 (2004–2005), p. 15 and p. 37.

[33] *Ibid.*, p. 15 and p. 27.

[34] *Ibid.*, p. 17. Cf. chapter 5 of the *Finnmark Act*, entitled "Surveying and recognition of existing rights".

[35] *Ibid.*, p. 17.

disagreements can be resolved through negotiations and consensus, which is reasoned in Sámi traditions.[36] Legal disputes arising from the process may be brought before the Land Tribunal.

Other significant changes in the draft Finnmark Act is that new Section 3, ensuring the commitments in international law, stating that the *Act* shall apply to the limitations imposed by the ILO Convention No. 169, and a section 10 that outline particular procedures including duties to hear the Sámi Parliament in cases of changes in the use of outlaying fields / uncultivated land and transfer of real estate property.

Further on, the Finnmark Estate is given a more independent possession than proposed in the Governmental bill, which prescribed that the Government should appoint a board member (without right to vote), is taken out of the act, and by the fact that the Finnmark Estate is given general expropriation protection.[37]

3 The Identification and Recognition Process of the *Finnmark Act* Chapter 5

3.1 The Mandate of Finnmark Commission

The mandate of the Finnmark Commission is given in the Fm Act S. 5, Para. 3, and is stated as to investigate rights to land and water in Finnmark "[i]n order to establish the scope and content of the rights held by Sámi and other people ...on the basis of prescription or immemorial usage or on some other basis".

In Section 29, this is specified to cover "rights of use and ownership to the land to be taken over [from the State] by the Finnmark Estate". It is also stated that the investigation shall be worked out "on the basis of to current national legislation". It is noteworthy that examination of reindeer husbandry rights, which is significant to the Sámi, is to be performed only upon demand by a person with a legal interest. The rights to salmonfishing in the large rivers of Finnmark, namely Tana and Neiden, are not included in the mandate of the Commission.[38]

Recently we have also, somewhat surprising, learned that the Finnmark Commission itself assumes that its mandate does not include determination of the indistinct boundaries between the lands of the Finnmark Estate and private properties measured before the Finnmark Act came into force (July 1, 2006).[39]

[36] *Ibid.*, p. 21.

[37] *Ibid.*, p. 15.

[38] Forskrift (regulation) 16. mars 2007 nr. 277om Finnmarkskommisjonen og Utmarksdomstolen for Finnmark, s. 5 and 6. An amendment of September 21, 2012 (no. 66), implies that the Finnmark Commission also has mandate to investigate claims of rights to fishing grounds in the coastal areas of Finnmark if someone with legal interest requeres it.

[39] See Finnmarkskommisjonen, *Rapport felt 1 Stjernøya/Seiland,* March 20, 2012 p. 15, where it is stated that there is "nothing in the Finnmark Act or the preparatory works that indicate that the legislator's intention has been that the Commission's reports should contain an accurate statement of the boundaries between properties and already meted and the Finnmark Estate".

In any case, the mandate of the Commission is far wider than comparable commissions, not limited to settle boundaries and disputes between the State and private lands, but aiming to investigate the full picture of rights that night exists on the former state lands of Finnmark.

The Commission is not a court of law. It is has therefore no mandate to settle judgments or other binding decisions. Instead, it has to prononce its findings as reports with legal conclutions, cf. Fm Act S. 33.

Establishing the Commission follows up on obligations to which Norway is bound by ILO Convention no. 169, particularly to identify indigenous lands and settle land claims under an adequate procedure within the national legal system, cf. Article 14 (2) and (3). The provisions aims to facilitate the identification process in relation to the Sámi, who, for the most part are locals living in villages or reindeer herders with winter residence in Inner Finnmark and the summer residence (and pastures) in the coastal areas. This holds not only for the formal process but also for the application of substantive law, including the use of legal sources as Sámi customary law, which I soon will return to.

Of importance to note is that the majority of the Standing Committee of Justice expressed great skepticism to the ordinary courts of law, stating that according to ILO Convention No. 169 article 14 (3), the scheme selected in the *Finnmark Act* was much preferable to the ordinary courts, where "it is clearly not acceptable under international law to hand over to the ordinary courts the question of which and the extent of rights acquired in Finnmark."[40]

As an additional argument for the proposed arrangement, the majority mentioned that there had been similar arrangements, regardless of indigenous peoples' rights and obligations under international law, elsewhere in the country.[41] The reasoning of the Standing Committee of Justice for proposing the identification process was thus in part due to Norway's international legal obligations to the Sámi, and in part that the people of Finnmark, Sámi and non-Sámi, should not be put in a worse position than people elsewhere in the country when it came to legal clarification of the status of outlaying fields and mountainous areas.

The Finnmark Commission consists of five members with a majority of lawyers with qualifications as judges (cf. Fm. act S. 29 para 2). The act does not set any requirement for Sámi or other representation except the fact that "at least two members shall be resident in or otherwise have a strong affiliation to the County of Finnmark". On the other hand, it is assumed that the Sámi Parliament is permitted to comment on the composition before the member is appointed by the government.[42] The Finnmark Commission was established by a Royal Decree of 14 March 2008

[40] *Innst. O. nr. 80 (2004–2005),* p. 28. This was later opposed by the Sámi Rights Committee II, see NOU 2007: 13 *Den nye sameretten,* p. 453. The Sámi Rights Committee II was appointed in 2001 to investigate the legal situation for the Sámi south of Finnmark. Their findings were published in NOU 2007: 13.

[41] Innst. O. nr. 80 (2004–2005) p. 28 where the Committee majority mention the Mountain Commission (1953–1953) and the Uncultivated Land Commission for Nordland and Troms (1985–2004).

[42] Innst. O. nr. 80 (2004–2005) p. 19.

pursuant to the *Finnmark Act*, S. 29, Para. 1 and the actual composition can be said to reflect a political balancing act with great emphasis placed upon correct ethno-political distribution with a substantial Sámi representation.[43]

3.2 The Considerations to Sámi Customs, Legal Opinions and Customary Law

In relation to the application of law, we have seen that the majority of the Standing Committee of Justice emphasized that the identification and recognition of rights should be based upon current *national law*. From the preparatory works it is shown that the term "national" was chosen instead of "Norwegian" to "better point out that consideration must be given to Sámi customs and legal opinions".[44] It shows that Sámi customs and customary law must be considered as substantive sources of law within the framework of ILO Convention no. 169, Article 8, and National Norwegian legislation.

Although the objective of this paper not is to analyze the weight of Sámi customary law in contradiction to Norwegian statutory law,[45] it does merit comment. Where indigenous people's customary law stands in conflict to other sources of law, the Sámi Rights Committee II has found that the weight of such law must be determined by the quality of the customs. The Committee does not preclude such customary law be given greater weight than customary law among the majority population, but rather concludes that "[c]ustomary law will not take unconditional precedence when in conflict with internal laws, nor in questions of law that do not apply fundamental legal principles."[46]

From the Norwegian Supreme Court verdict published in Norsk Retstidende [NRt.] 2001 p. 1116, it is stated that Sámi customs had to be clear and have a certain quality.[47] Two prejudicing cases, the Selbu and the Svartskog, published in NRt. 2001 p. 759 and NRt. 2001 p. 1229, respectively, are important sources when Sámi land rights are to be clarified. The majority of the Standing Committee stated that:

---

[43] See http://www.domstol.no/Enkelt-domstol/Finnmarkskommisjonen/Om-kommisjonen/Medlemmer-og-ansatte/. Retrieved May 15, 2011, where the composition of the Commission is shown. Two members are Sámi from the Sami areas of Inner Finnmark, one is from the coastal areas and one is representing the outdoor interests. The head, Jon Gauslaa was formerly the head of the Sámi Rights Committee II. He is from southern Norway and has a most "neutral" background.

[44] Innst. O. nr. 80 (2004–2005), p. 19.

[45] For further reading, see Øyvind Ravna, 'Sámi Legal Culture – and its Place in Norwegian Law in *Rendezvous of European legal cultures,* eds. Jørn Øyrehagen Sunde and Knut Einar Skodvin, Bergen 2010, pp. 149–165.

[46] NOU 2007: 13, p. 222.

[47] See also case law published in Nrt. 2008 p. 1789 on the evaluation Sámi customary law.

> Assessment of evidence in the recent case law has been satisfactory. Modern Norwegian case law, particularly the Selbu and Svartskog cases, has given instruction on how traditional Sámi use shall be considered as a basis for acquisition. These will be important sources of law for the Commission and Court.[48]

The Committee actually went so far as to discuss whether this "recent case law" should be codified in the *Finnmark Act*, but did not propose it since it would mean that statutory provisions and not case law would be the sources in the identification process. This shows, however, that these cases represent important sources of law in answering substantive questions about when rights are to be identified in Sámi areas.

Finally, the commitment has been strengthen through Norway's signing of the *UN Declaration on the Rights of Indigenous Peoples*, which in Article 40 states that the settlement of disputes relating to indigenous peoples shall take into account "the customs, traditions, rules and legal systems of the indigenous people's concerned and international human rights." This provision can be compared with the ILO Convention No 169 Article 8. The UNDRIP Article 26, which ensures the Indigenous Peoples' right to own, develop and control the lands, territories and resources that they possesses, is for sure also of significance.

3.3 Some Other Procedural Law Requirements of Importance

The process for the Finnmark Commission does not begin with a claim, a suit or other party subpoena, like the case is for ordinary courts of law or was for the Uncultivated Land Commission of Nordland and Troms. The Finnmark Commission is neither assigned investigation fields by central authorities, as was the Mountain Commission working in the southern mountainous areas (1908–1953), but shall *itself* determine which fields it will investigate and the sequence of the hearings, cf. Fm. act S. 30, Para. 1.

Section 30 also states that consideration shall be placed on "natural and appropriate delimitation of the field as regards extent and legal and historical context and the need to clarify the legal relations." Based on experience from the first three fields which have been taken for investigation,[49] it can be said that the Commission has placed greater emphasis on natural and appropriate delimitation rather than the need for clarification. By the selection of field 4, Karasjok / Kárášjohka (opened for investigation January 25, 2011, and not completed by October 2015), the Commission has chosen to investigate one of the most demanding Sámi areas where it is a great need for internal legal clarification of among others the reindeer husbandry rights and areas.

[48] Innst. O. nr. 80 (2004–2005), p. 36. More about the cases, see Øyvind Ravna, 'The Process of Identifying Land Rights in parts of Northern Norway' pp. 429–432 and Gunnar Eriksen, *Alders tids bruk,* Bergen 2008 pp. 324–348.

[49] The three first fields of the Commission is Stierdna-Sievju/Stjernøya-Seiland (field 1), Unjarga/Nesseby (field 2), and Sállan /Sørøya (field 3), cf supra note 8. Fields 1 and 3 consist of islands in the Alta Fjord in West-Finnmark, while field 2 is a municipality in eastern Finnmark.

The Commission can omit investigation consideration of cases "that are clearly inappropriate for investigation by the Commission", cf. Fm. act S. 30, Para 3. For such a decision, emphasis should be put on the character of the right and its legal basis. The majority of the Standing Committee of Justice has pointed out that such assessment must be based on the background and purpose of the identification process, where

> a right based on immemorial usage normally will fit better than a right based in a contract. Uncultivated areas will normally be better suited for investigation than a right to rent or lease ground.[50]

It is further stated that the Commission primarily shall investigate rights of use and ownership that are based on long-term and traditional use. But the mandate cannot be limited to this: "According to the majority's opinion, it is therefore difficult in a precise way to specify in more detail the legal questions the Commission shall examine."[51]

The Finnmark Commission has the responsibility for case illumination, cf. Fm. act S. 32 Para 1, which is natural since it is an investigatory body, and not a court of law. The act further states that

> the Commission may in the manner it finds appropriate obtain statements, documents and other material and conduct surveys and investigations, etc. concerning actual and legal circumstances that may be significant for the Commission's conclusions.[52]

However, the Finnmark Act does not prevent the parties themselves from illuminate the facts or the evidence for the Commission. Representatives for interest groups may also be appointed to follow the working of the Commission. The cost shall be covered by the state, cf. Fm. act S. 32, Para 3. But opposite the previous judging commissions, the state doesn't provide the parties cost for legal counsels.

As mentioned, the Finnmark Commission is not a Court of Law and is thus not going to file a judgment. The findings of the Commission shall be submitted in a report on the legal status of the investigation field. The report has to contain information about (a) who, in the view of the Commission, are owners of the land, (b) what rights of use exist, and (c) the circumstances on which the Commission bases its conclusions, cf. Fm. act S. 33. The Commission cannot refuse to consider an ownership dispute, for example, by concluding that it is other than the Finnmark Estate who is the owner of a particular piece of land. The majority of the Standing Committee of Justice here points out that the Commission in such cases

> must take a standing on what result has the best basis in law. It is not acceptable to only conclude that the Finnmark Estate is not the owner of the area in question without also indicating who is assumed to be the owner.[53]

To my point of view, this means that the Commission is committed to investigate and propose lines for unclear boundaries between the lands of The Finnmark Estate

---

[50] Innst. O. nr. 80 (2004–2005), p. 19.

[51] *Ibid.*, p. 20.

[52] The Finnmark Act S. 32 Para 1, 2nd sentence.

[53] *Ibid.*, p. 21.

and private parties, too. Elsewhere it has not completed the identifying of the rights on the lands of the Finnmark Estate.

The Commission's reports will in not have legal efficacy. Legal effect depends on agreements between the parties, unilateral declarations or that the case is brought further to the Land Tribunal.[54]

3.4 *The Duties of the Finnmark Estate and the Private Parties*

The Finnmark Estate is mandated to without undue delay assess the conclusions of the Finnmark Commission, cf. Fm. Act S. 34 Para 1. This is natural and necessary since the Finnmark Estate holds the title to the lands examined by the Commission and is thus landowner and party to the investigation. The Standing Committee of Justice has pronounced that the Finnmark Estate is more than an ordinary landowner and party, and has commitments in the identification process beyond what can be termed as ordinary party obligations.[55]

The Finnmark Estate has thus obligations to actively consider the Commission's report. To the extent the Finnmark Estate agrees that others have rights on the land presently owned by the Finnmark Estate, it is obliged to confirm and without undue delay attend the rights to be registered. Through agreement, negotiated consensus or unilateral declaration, the process will terminate at this stage, cf. Fm. act S. 34 Para. 2.

Private parties have not such obligation to act on the report of the Commission. The majority of the Standing Committee of Justice in practice assumed the opposite, when it stated "if the parties do not want to put the question to tip by bringing it to the Land Tribunal, they can allow the Finnmark Estate to continue to manage the grounds without cutting off the possibility of raising the issue at some future point in time."[56] That statement virtually proposes leaving the legal issues *unresolved* without legal efficiency, and nearly sustain parties in waiting to put forward a claim for strategic reasons. Such reasons can be assuming that prescription period is not yet reached; future change in the interpretations of the law gives better possibilities, or other circumstances that may later work to one's benefit. Likewise the reason not to respond might be kind of legal insecurity between a dispute and agreement.

According to the Fm. act S. 35, parties that do not agree with the Commission's conclusions or who need assistance to have the conclusions transferred to a binding agreement with the opposite party, can ask the Finnmark Commission for mediation after the report is filed.

It is also notable that disputes are assumed to be resolved according to Sámi tradition. The majority of the Standing Committee of Justice here refers to the Sámi Parliament, which has emphasized that "conflicts as far as possible and in line with

[54] For more reading (in Norwegian), see Øyvind Ravna, 'Rettsvirkningen av rettskartleggings og anerkjennelsesprosessen i Finnmark', *Lov og Rett,* (50) 4/2011, pp. 220–240.

[55] *Ibid.,* p. 21.

[56] *Ibid.,* p. 21.

Sámi traditions shall be resolved through negotiations and not through court proceedings."[57] The majority stated that it support such a procedural approach completely.

3.5 The Land Tribunal for Finnmark

The Land Tribunal for Finnmark has a mandate to hear "disputes concerning rights that arise after the Finnmark Commission has investigated a field", cf. Fm. Act S. 36, Para 1.[58] The Tribunal shall consist of five members, where the chair, the vice-chair and one other member shall fulfill the requirements for Supreme Court judges. There is no demand for any other requirements as locals or other special knowledge by the members, not even a connection to the County of Finnmark.

General civil procedural rules apply in the same way for the Land Tribunal for Finnmark as they did for previous comparable tribunals as the Uncultivated Land Commission for Nordland and Troms, so far as they are applicable, and nothing else is specified in the act, cf. S. 46, Para. 2. But like the former tribunals, there are a number of special procedural provisions. As we have already seen, there are particular rules on arbitration, where the Finnmark Commission is given a duty in mediation. However, the mediation is not compulsory,[59] which means that legal proceedings can take place once the Commission has submitted its report.[60] The Land Tribunal itself is not assumed to carry out court mediation or other forms of mediation.

A period of one year and six months is set to bring disputes that arise after the Finnmark Commission has investigated a field, before the Tribunal. The period runs from the time the Commission has submitted its report, cf. S. 38, Para. 1. The deadline is assumed to be long enough to allow the Finnmark Estate time to consider the report and to give the parties' time to negotiate. The extended period of time can also be explained in that "regards to some Sámi ways of utility also implies a need for a long period",[61] presumably supposing that what is left of the Sámi nomadic livelihood, needs a longer time to respond. The majority also argues that the long period of time can contribute in impelling the negotiations forward and put pressure on the parties to reach consensus.

But the extended deadline in bringing the dispute before the Land tribunal may be problematic in relation to the requirement for trial within a reasonable time in relation to the provision in the *European Convention on Human Rights* (ECHR)

[57] *Ibid.*, p. 21.

[58] The tribunal must also be seen as part of Norway's obligation under Article 14 of the ILO Convention No. 169. The Tribunal is not yet established (May 30, 2012).

[59] Innst. O. nr. 80 (2004–2005), p. 22.

[60] At the moment this is however problematic, since the first report was filed March 20, 2012, and the Tribunal is not yet established (May 30, 2012).

[61] Innst. O. nr. 80 (2004–2005), p. 23.

Article 6.[62] It is not discussed in the preparatory works, except for the statement of the Ministry of Justice saying that one of the aims of the Finnmark Act is that "the legal situation throughout Finnmark will be investigated within a reasonable time".[63]

The long deadline is neither not exhaustive. The Land Tribunal may deal with matters that come in at a later stage if not all cases in a field have been brought to conclusion and if it finds the case appropriate for such consideration, cf. S. 38, Para 2.

The Land Tribunal has exclusive jurisdiction, cf. S. 36, Para 3, which means that cases that fall under the Tribunal cannot be brought before the ordinary courts or the land consolidation courts except in specified circumstances. Such circumstances occur when the Tribunal has dismissed a case pursuant to Section 39 (see below) or if the deadline for bringing proceedings before the Tribunal has expired. The exclusive competence means that *lis pendens* in a certain investigation field occurs when the deadline for bringing the matter before the Tribunal starts to run, i.e. after the Commission has submitted its report.[64] In fact exclusive competence will block lawsuits by the ordinary courts until the last dispute in an investigation field is processed.

Questions that are "found inappropriate for consideration" by the Tribunal may be dismissed in whole or in part, cf. S. 39, Para. 1. Such rejection can be done *ex officio* and cannot be appealed, cf. S. 39, Para 2. The claimant, however, shall be allowed to respond before dismissal occurs. When it comes to matters or disputes that are not suitable for treatment, it is comparable to those the Finnmark Commission can refuse to investigate, pursuant to Fm. act S. 30, Para 3. The threshold for rejecting a claim is in any case somewhat lower, since it is not required "that the case is obviously not suitable for treatment".[65]

Although an appeal cannot be posed against rejection of such Court rulings, the Majority of the Standing Committee of Justice assumes that the interests of the plaintiff are met since he is allowed to respond before the Tribunal rejects the question. The majority also states that the Tribunal should

> be able to concentrate on the major and fundamental issues, so that minor matters, such as … adjusting the boundaries between two properties, or interpretation of contracts for the sale of property, could be left to the ordinary courts or land consolidation courts.

The substantive decisions; the judgments of the Land Tribunal, can only be appealed directly to the Norwegian Supreme Court of Justice, cf. Fm. act S. 42. The Majority of the Standing Committee of Justice points out that a "similar solution was selected for the Uncultivated Land Commission for Nordland and Troms."[66]

[62] See Øyvind Ravna, 'The Finnmark Act 2005 Clarification Process and Trial 'Within a Reasonable Time", *Nordic Journal of Human Rights* (29) 2/2011 pp. 184–205.

[63] Note to Section 2 of the Regulation on the Finnmark Commission (March 16, 2007 No. 277) of Royal Decree 16 March 2007 p. 3.

[64] Innst. O. nr. 80 (2004–2005), p. 23 where it is stated "[i]f the deadline is exceeded, a party may bring the matter before the ordinary courts".

[65] *Ibid.*, p. 23.

[66] *Ibid.*, p. 25.

But The Majority did not consider that these rules of appeal, which was originally adopted for the Mountain Commission in 1908, were severely limited because of an amended to Civil Procedure Act of 1915 by Act 22 May 1981 No. 24.[67]

The chosen appeal procedure means that presumably only a minority of the appeals will be heard, since the Supreme Court is not an ordinary court of appeal, but rather a Court for settling principle questions of broad significance outside the concerned parties. Another objection against this appeal scheme is that the Norwegian Supreme Court can neither make on-site inspections nor examine witnesses itself.

4 Final Remarks

In reviewing the identification and recognition process of the Finnmark Act, one can say that the Act is both innovative and unique, not only because of the influence of the indigenous people in the legislative process, but because it aims to take into account the commitments in the ILO Convention no. 169 regarding identification of Sámi ownership, the use of Sámi Customary law, other customs and traditions, including a the Sámi particular way of life. The construction with two independent bodies put together in a unified system, may also be considered as an innovation.

We may note that the Sámi Rights Committee II, primarily based on review of case law of the ILO monitoring bodies,[68] have concluded

> that the Finnmark Act system as a whole clearly must be considered to meet the requirements of ILO Convention No. 169 Article 14 (2) and 14 (3).[69]

It also points out that the solution chosen for Finnmark must be considered to be "in line with the Norwegian aims to loyal achieve the purpose of the ILO Convention" and thus as an adequate procedure within the national legal system to resolve land claims from the Sámi.

But the combination of an investigation body and a special land tribunal is also challenging, particularly in respect to more practical approaches. Such question do not seem to be reviewed to the same extend by the legislators or the Sámi Rights Committee II.

The analysis shows that it is relevant to quarry if procedural requirements in the Act, like the upheld of the party disposal, the opportunity to achieve decisions with

---

[67] The amended to the civil procedure Act 22 May 1981 No. 24 meant that the access of appeal was strictly limited in relation to previous Commissions, where the judgments could be appealed directly to the Supreme Court. The amendment had its origin in that the Supreme Court in 1979 proposed to limit the appeal right so that the Appeal Committee of the Supreme Court can refuse to promote the case under the Civil Procedure Act S. 373, para 3 (4), if it found "that neither the decision importance beyond this case or other circumstances give reason that the appeal will be tried by the Supreme Court".

[68] See NOU 2007: 13 pp. 431–455.

[69] *Ibid.*, p. 453.

legal efficacy, and access to efficient appeal remedies. The two-body identifying clerifying and dispute solving procedure process also involves a challenge to the rules of trial within a reasonable time in the European Convention on Human Rights. Effective remedy of appeal and trial within a reasonable time can be seen as conflicting interests, but both these requirements must be met according to ECHR Article 6.[70] That the Commission has chosen to define the determination of boundaries out of its mandate means that the practical significance and implication of the investigation work is reduced.

In relation to the substantive side of the law, the procedure may be challenging, too, not only because the Finnmark Commission and Land Tribunal have a far wider mandate than comparable former commissions, but also because they have less guidance from preparatory works and case law than the comparable commissions have had. That the process occurs in a part of the country, or in a culture, where property law traditions have lower standing than elsewhere, also adds to the difficulty.[71] And even if there are some land marking cases,[72] case law indicates that it will take time to establish norms for clarifying and ensuring the quality of Sámi customary law.[73]

Sámi customary law is however an important source of law, not only because of the ILO Convention no. 169 and the place such sources is given in the preparatory work of the Finnmark Act, but also due to the allowances to Sámi legal culture we must expect from Norwegian legal culture.

Since the Finnmark Act overall has some constraints, few sources and precedents to depend upon, one problem may be the predictability of a case outcome or an answer to a legal question. Consequently, it can be difficult for the claimant or parties to predict the result of a particular case. While the previous Uncultivated Land Commissions for Nordland and Troms only had to investigate whether the state owned the land, the boundaries between state and private land, and what rights of use existed on the land belonging to the state, the Finnmark Commission has to examine the whole bundle of rights and resources that might be found on what today is the Finnmark Estate. It might include community commons, joint-ownership and Sámi siidas.

Even if the current Finnmark Commission is situated with Sámi members, there is no requirement for local knowledge or knowledge of Sámi customs and customary laws, neither among the members of the Commission nor among the members of the Tribunal. It may be problematic. Local knowledge is generally important for reaching a correct and reasonable result *de facto* acceptable to all parties. It is also important for parties to know that their peers have contributed to the decision. Sámi customs, customary law and legal traditions are little-taught in law schools today, so *Sámi* local knowledge is therefore, of paramount importance.

---

[70] See also the ILO Convention No 169 Article 12 and 14 (3) and the UN Declaration on the Rights of Indigenous Peoples Article 40.

[71] Comparable arrangements in other countries could be discussed, but there is probably not much to be mentioned. For an overview, see NOU 2007: 13, pp. 247–271.

[72] See NRt 2001 p. 769 (Selbu) and NRt. 2001 p. 1229 (Svartskogen), which set up norms for how the rules on immemorial usage are to be applied to Sámi land claims.

[73] See NRt. 2001p. 1116 and NRt. 2008 p. 1789.

Space does not permit a *de lege ferenda* discussion (how the law ought to be). However, the review shows that people in Finnmark, both Sámi and non-Sámi, probably will gain from some amendments to the law. To ensure trial within a reasonable time, there could be sat a shorter period of time to bring cases before the Tribunal. Transferring the Commission to a *court of law* should be given considerations, too, as it would benefit both to a more predictable remedy of appeal, less proceeding time, and provide the opportunity to obtain enforceable decisions. With such amendments, cases should naturally start with a writ or a lawsuit, with the Finnmark Estate and those who appoint its board, the Sámi Parliament, and Finnmark County Council, playing an active role. Both the Finnmark Commission and Tribunal should then by law be ensured a larger proportion of qualified lay persons, especially with local knowledge and understanding of Sámi customary law.

Finally, and in spite of the infirmity pointed at above, I will like to express that the review shows that Norway as a state is *recognizing* Sámi rights to land and natural resources, giving the Sámi representatives a rather substantial influence in the legislative process of the Finnmark Act. The Act can thus be uphold as an example for the management and self-determination of the natural resources in a core Sámi area. As mention in the introduction, this was pointed out by the UN's Special Report on Indigenous Peoples' Rights, James Anaya in 2011,[74] who also uphold the Act as an important example for the other Nordic Countries.

At the same time the UN Rapporteur expresses certain reservations, stating that since "the process for identifying rights to land under the Finnmark Act is currently underway, the adequacy of the established procedure is not yet known."[75]

Such a reservation is relevant. But it can hardly be addressed to a lack of upholding the commitments of ILO Convention no. 169 or failures in the legislative process, but rather that the application of law by the Finnmark Commission wasn't known at that time. Therefore, it is also fair not to conclude on the process is an important example for other countries with indigenous people before more of the result of the Finnmark Commission is revealed.[76]

However, there are there might be failures explained from uncoordinated and inadequate preparatory work, too. This means that even if people in Finnmark; Sámi and non-Sámi, can subsist with the current Act, there is considerable room for improvement, where it is possible to establish a more adequate procedures within the national legal system to resolve land claims by the Sámi. This may not primarily

[74] James Anaya, *The situation of the Sámi people in the Sápmi region of Norway, Sweden and Finland.* Retrieved September 20, 2011, from http://unsr.jamesanaya.org/docs/countries/2011_report_Sámi_advance_version_en.pdf, paragraph 44.

[75] *Ibid.,* Para. 49.

[76] The reservation has been more relevant as the Finnmark Commission has delivered more investigations (see supra note 8); in the first three reports there are not found collective use rights beyond the extend of the Finnmark Act and no ownership rights of the Sámi or non-Sámi locals. See also the additional remarks of the author on the bottom of the main text. For more reading on the Finnmark Commission's findings, see Ravna, Ø. (2013). The First Investigation Report of the Norwegian Finnmark Commission, International Journal on Minority and Group Rights, 3(20), 443–457 and (in Norwegian) Ravna, Ø. (2013), Finnmarkskommisjonens bevisvurderinger og rettsanvendelse – drøftet ut fra dens to første rapporter, Lov og rett, 8 (52), 612–631.

be reasoned to better meet the requirements of the ILO Convention, but to fulfill the more practical demands like legal efficacy, party disposals and consume of time. Such improvement is a responsibility for the legislature, in cooperation with the Sámi representatives, which are to participate according to the principle of free, prior and informed consent, based on the international legal commitments, constitutional obligations and the moral obligations of a State that possess territory of an indigenous nation.

After this text was submitted in 2012, the Finnmark Commission has completed three more fields of investigation; totally four fields. The results of these investigations show that the reservation the UN Raporteur of Rights of Indigenous Peoples expressed were hightly relevant. In none of the four fields investigated, has the Finnmark Commission identified any land collectived owned by the Sámi. Neither has the Commission found any use rights of such nature that the Sámi have access to dispose their rights, regulate or control the use, or benefit from the usufructs of these. If the procedure Norway has chosen to comply with the ILO Convention no. 169 shall be considered as adequate within the national legal system, both in realtion to identify Sámi lands, protecting it or to solve claims, as precreibed in that convention, the result of the forthcoming investigations have to be quite different from the first four.

Global Context – Arctic Importance: Free, Prior and Informed Consent, a New Paradigm in International Law Related to Indigenous Peoples

Leena Heinämäki

Abstract In international law, a fundamental shift is currently occurring in State-Indigenous relations, which can be seen as culminating in the adoption of the UN Declaration on the Rights of Indigenous Peoples, and at its endorsement of the right of indigenous peoples to self-determination and a free, prior and informed consent (FPIC) in the decisions that concern them.

This chapter aims to study and analyze FPIC in the light of multiple important developments that have prepared and pushed states to slowly accept that indigenous peoples cannot be seen merely as objects of protection but must be recognized as serious actors and as "partners" with and within the nation states. When implemented, the right to FPIC has positive effects on the important issues such as indigenous peoples' land use and governance. This is of a particular importance in the Arctic that is the homeland for a great number of indigenous peoples.

The article starts with a general overview over the past decades, to show how a paradigm shift has emerged and is currently evolving in international law concerning the rights and the status of indigenous peoples, in order to understand and give weight to the concept and the right of FPIC. After this overview, setting a necessary context to FPIC, this chapter moves on to take a closer look at the right of FPIC itself, and how it is articulated in indigenous rights instruments as well as in the case law and observations of human rights monitoring bodies. Examples are given from Arctic areas when merited. Examples from other areas, however, are not less valid should the same circumstance prevail in one of the Arctic countries, since international law naturally binds all the states.

Keywords Indigenous peoples • Human rights • Free Prior Informed Consent • Right to participation • Arctic

L. Heinämäki (✉)
Northern Institute for Environmental and Minority Law, University of Lapland, Rovaniemi, Finland

Arctic Centre, Rovaniemi, Finland
e-mail: leena.heinamaki@ulapland.fi

T.M. Herrmann, T. Martin (eds.), *Indigenous Peoples' Governance of Land and Protected Territories in the Arctic*, DOI 10.1007/978-3-319-25035-9_11

1 Introduction

Despite the fact that international law can justifiable be criticized for its many deficiencies, it should be noted that in relation to indigenous peoples' international status, quite profound developments have taken place during the last couple of decades, especially the last few years. It can convincingly be argued that in international level, a fundamental shift is currently occurring in State-Indigenous relations, although this shift does not yet fully reflect in the national or local implementation. This shift, however, can be seen as culminating in the adoption of the UN Declaration on the Rights of Indigenous Peoples,[1] and at its endorsement of the right of indigenous peoples to self-determination and a free, prior and informed consent (FPIC) in the decisions that concern them.

This chapter aims to study and analyze FPIC in the light of multiple important developments that have prepared and pushed states to slowly accept that indigenous peoples cannot be seen merely as objects of protection but must be recognized as serious actors and as "partners" with and within the nation states. When implemented, the right to FPIC has positive effects on the important issues such as indigenous peoples' land use and governance. This is of a particular importance in the Arctic that is the homeland for a great number of indigenous peoples. Since this chapter focuses on developments of international law, applicable throughout the world, it does not have an explicit Arctic focus. Arctic states, however, are bound by the discussed international norms. Additionally, most of the Arctic states are being active in their human rights policies, also regarding indigenous peoples. Finally, it should be mentioned that the Arctic Council, an inter-governmental organization created by the Arctic States to enhance environmental protection, has prepared the governments and officials of the Arctic states to work closely with indigenous peoples' organizations that enjoy a status of permanent participants in the Council.

This chapter starts with a general overview over the past decades, to show how a paradigm shift has emerged and is currently evolving in international law concerning the rights and the status of indigenous peoples, in order to understand and give weight to the concept and the right of FPIC. After this overview, setting a necessary context to FPIC, this chapter moves on to take a closer look at the right of FPIC itself, and how it is articulated in indigenous rights instruments as well as in the case law and observations of human rights monitoring bodies. Examples are given from Arctic areas when merited. Examples from other areas, however, are not less valid should the same circumstance prevail in one of the Arctic countries, since international law naturally binds all the states.

---

[1] The United Nations Declaration on the Rights of Indigenous Peoples, September 7, 2007, Sixty-first Session, A/61/L.67.

2 The Context: Paradigm Shift in Indigenous Peoples' International Status

From the 1970s onwards, indigenous peoples' organizations become active in relation to UN human rights bodies where they started to participate with increasing frequency, demanding their human rights to be recognized and implemented. Indigenous peoples have enhanced their access to these bodies as several indigenous peoples' organizations have achieved official consultative status with the UN Economic and Social Council. Indigenous peoples also have approached regional human rights bodies, particularly Inter-American Commission on Human Rights, with successful outcomes, as will be described later.[2]

An important development regarding the status and rights of indigenous peoples started in 1982 by the establishment of the Working Group on Indigenous Populations (WGIP) under the United Nations Economic and Social Council.[3] The WGIP's original mandate was to review developments concerning indigenous peoples and to work toward the development of corresponding international standards.[4] Additionally, however, through its policy of open participation in its annual sessions, the WGIP became an important arena for the dissemination of information and exchange of views among indigenous peoples, governments, non-governmental organizations, and others.[5] WGIP played a significant role in strengthening the status of indigenous peoples as they were accepted as actors to contribute developments towards an international recognition of their rights.

In 1985 the Sub-Commission on Prevention of Discrimination and Protection of Minorities approved the WGIP's initiative to draft a universal declaration on the rights of indigenous people for adoption by the UN General Assembly.[6] After many phases, in 1994 the sub-commission adopted the working group's draft and submitted it to the UN Commission on Human Rights, which subsequently established its own ad hoc working group to work on the declaration.[7]

Through the process of drafting a declaration, WGIP engaged states, indigenous peoples and others in a broad multilateral dialogue on the specific content of norms concerning indigenous peoples and their rights. This was a very important step historically, since the working group provided an important means for indigenous peoples to promote, for the first time, their own conceptions of their rights within

[2] *Ibid.*

[3] Economic and Social Council Resolution 1982/34, Retrieved January 19, 2014, from http://ap.ohchr.org/documents/E/ECOSOC/resolutions/E-RES-1982-34.doc. The WGIP is an organ of the Sub-Commission on the Promotion and Protection of Human Rights.

[4] Human Rights Commission Res. 1982/19 (Mar. 10, 1982); E.S.C. Res. 1982/34, May 7, 1982, U.N. ESCOR, 1982, Supp. No. 1, at 26, UN Doc. E/1992/82 (1982), paras. 1–2.

[5] Anaya (2004), Indigenous Peoples in International Law, 2nd edn, Oxford, at 63.

[6] Sub-Commission on Prevention of Discrimination and Protection of Minorities Res. 1985/22 (29 August 1985).

[7] Commission on Human Rights, Resolution 1995/32 (3 March 1995).

the international arena, enabling them to make proposals and comments.[8] Eventually a great number of states and other actors came to participate in the working group discussion on the declaration,[9] which became finally adopted as the UN Declaration on the Rights of Indigenous Peoples two decades later in 2007. It is important to note that the very reason for the process to take more than two decades is the persistent push from indigenous delegates to have the most crucial yet the most controversial paragraphs, relating to self-determination and FPIC, accepted by the states. Indigenous peoples were able to shake the very basis of the human rights structure, focusing on the individual rights, and push states to accept their collective rights.

The above mentioned developments have had many profound effects also on legal practice and statements of the human rights monitoring bodies, culminating in the recent acknowledgment of the right to FPIC. When the International Covenant on Civil and Political Rights (ICCPR)[10] was adopted in 1966 (entering into force 1976), the original purpose of its minority Article 27 was not aiming to create any special rights for indigenous peoples, but to make sure that minorities were not denied by the states to enjoy their culture in a peaceful way. Despite the negative expression in Article 27: "minorities shall not be denied to enjoy their culture," the UN Human Rights Committee (HRC), the monitoring body of ICCPR, has started to derive positive obligations from this provision.[11] In its jurisprudence concerning indigenous peoples, the HRC has emphasized two requirements: consultation and economic sustainability of the traditional, nature-based livelihood as an inherent part of indigenous culture.[12]

A close connection of indigenous peoples to their traditional lands and natural resources is the very reason for rather extensive developments regarding their international status that has exceeded the general status of minorities in international law. According to the new paradigm, indigenous peoples can be seen as "semi-subjects" of international law, which do not enjoy only quite extensive substantive protection, but whose procedural rights, particularly participatory rights, have been developed beyond the participatory rights of other minorities or the general public.

The HRC, due to its recognition of an intimate link between indigenous culture and natural resources, has started to apply Article 1 of ICCPR (the right to self-

---

[8]Anaya (2004), *supra* note 5, at 63–64.

[9]Anaya (2004), *supra* note 5, at 64.

[10]The International Covenant on Civil and Political Rights, adopted 14 December 1966, entered into force 23 March 1976, 999 UNTS 302.

[11]*Lubicon Lake Band v. Canada*, Communication No. 167/1984, CCPR/C/38/D/167/1884. UN Human Rights Committee, General Comments No. 23: The Right of Minorities (Art. 27), UN Doc. CCPR/C/21/Rev.1/Add.5,8 April 1994, Para 6.1.

[12]*I Länsman et al v. Finland*, Communication No. 511/1992, UN Doc.CCPR/C/52/D/511/1992 (1994). For the two-part test of consultation and economic sustainability, See Scheinin, M. (2000). The Right to Enjoy a Distinct Culture: Indigenous Land and Competing Uses of Land. In Orlin, T.S., Rosas A. and Scheinin, M (Eds), The Jurisprudence of Human Rights Law: A Comparative Interpretive Approach (pp. 159–222, at 168). Institute for Human Rights, ÅboAkademi University, Turku/Åbo.

determination of peoples) to indigenous peoples in its Concluding Observations related to the mandatory reports of the State Parties. This shift originally took place in 1999, as a part of many other developments and active lobbying related to the rights of indigenous peoples. The recognition of the applicability of the right to self-determination on indigenous peoples has been followed by other human rights monitoring bodies, such as the Committee on Economic, Social and Cultural Rights (CESCR), the Committee on the Elimination of Racial Discrimination (CERD) and the Inter-American Human Rights Court.[13]

During 1960–1970, assimilation policies, with the aim to integrate indigenous peoples to the mainstream population without recognising their right to specific cultural characteristics, had to make space for demands of indigenous peoples' movements concerning their right to self-determination, self-government and the positive recognition of their cultural integrity. The first International Labour Organization's (ILO) Convention on Indigenous and Tribal Populations of 1957[14] with the strong integration approach was replaced by the ILO Convention on indigenous peoples No. 169 of 1989,[15] marking a change not only in ILO's approach to indigenous peoples but also is reflecting the shift in the general attitude towards indigenous peoples. The latter Convention is based on respect for indigenous peoples' specific cultures and distinct ways of life, and their traditional forms of governance and their customs and own legal systems.

ILO Convention No. 169 contains not merely substantive protection of the cultural integrity of indigenous peoples but, as reminded by Barsh, also recognizes them as political and legal entities that need to be taken into account in the decision-making in the matters that concern the group.[16] The clear aim of the convention is to establish a frame for the partnership between indigenous peoples and the state governments. The equal partnership-idea in ILO Convention No. 169, however, remains short of full recognition: first of all, indigenous peoples themselves were not able to participate in the creation of the Convention. Secondly, the Convention does not recognize the rights to self-determination and self-governance of indigenous peoples, which means that it still regards indigenous peoples as objects of protection rather than active subjects of co-operation.

Yet, ILO Convention No. 169, although ratified only by 22 countries, has played a significant role in the improvement of the international status of indigenous peoples. The ILO Convention No. 169 is the only international treaty solely concerned

[13] *Saramaka v. Suriname*, Inter-American Court of Human Rights, Judgment of November 28, 2007, Series C, No 172, Para 93.

[14] ILO Convention No 107 Concerning the Protection and Integration of Indigenous and other Tribal and Semi-Tribal Populations in Independent Countries, adopted 26 June 1957, entered into force 2 June 1959, 328 UNTS 247.

[15] International Labour Organization (ILO) Convention (No. 169) concerning Indigenous and Tribal Peoples in Independent Countries, Geneva, adopted 27 June 1989, entered into force 5 September 1991, 28 ILM 1382.

[16] Barsh, R.L. (1994). Indigenous Peoples in the 1990's: From Object to Subject of International Law? Harvard Human Rights Journal, 33, reprinted In Watters, L. (2004) (Ed.), Indigenous Peoples, the Environment and Law (pp. 15–42, at 23). Carolina Academic Press, North Carolina.

with indigenous peoples. It can be seen significant to the extent that it creates treaty obligations among ratifying states in line with current trends in thinking prompted by indigenous peoples' demands. As reminded by Anaya, the Convention is important as part of a larger body of developments that can be understood as giving rise to a new customary international law with the same normative thrust.[17]

Indigenous peoples' legal status has been evolved also as a part of the concept of sustainable development as well as biodiversity protection regime. In 1992, Rio Conference on Sustainable Development (UNCED) shifted the role of indigenous peoples from the objects of protection to subjects of co-operation. The outcomes of the Conference created an active role for indigenous peoples by launching the idea of partnership between states and indigenous peoples in the maintenance of the sustainability and the protection of the world's biodiversity. Principle 22 of the Rio Declaration[18] recognizes indigenous peoples as distinct social partners in achieving sustainable development, emphasizing the unique value of indigenous traditional cultures and ways of life.[19]

In a similar way, the Convention on Biological Diversity (CBD) establishes an agency for indigenous communities to participate in the maintaining of environmental sustainability. Article 8 (j) is the key provision of the CBD in relation to indigenous and local communities. It has had a considerable impact on international discussions related to the status of indigenous peoples, including the advancement of the FPIC.[20] The CBD furthermore strengthens the recognition of the concept of "benefit sharing",[21] which was already acknowledged by ILO Convention No. 169,[22] indicating that indigenous peoples have to be treated as partners with sharing the benefits of the projects that take place in their traditionally used lands.

The term "partnership" was used in Agenda 21. Since Agenda 21 was adopted by consensus at the UNCED, states could not object to the use of "partnership" in other United Nations contexts.[23] By the time of the UNCED preparatory negotiations in

[17] Anaya, J. (2005). Indigenous Peoples' Participatory Rights in Relation to Decisions About Natural Resource Extraction: The More Fundamental Issues of What Rights Indigenous Peoples Have in Land and Resources. Arizona Journal of International & Comparative Law, 22(1), at 9.

[18] The Rio Declaration on the Environment and Development, United Nations Conference on Environment and Development, Annex, Resolution 1, UN Doc. A/Conf.151/26/Rev.1 (vol1), (1992), 31 ILM 874.

[19] Barsh, R.L. (2004), *supra* note 16, at 23.

[20] Stoll, P.-T. & von Hahn, A. (2005). Indigenous Peoples, Indigenous Knowledge and Indigenous Resources in International Law. In Ghanea, N. & Xanthaki A. (Eds), Minorities, Peoples and Self-determination (pp. 3–14, at 33). Dordrecht: Martinus Nijhoff Publishers. The Nagoya Protocol (to CBD) of 2010 recognizes and advances the concept of FPIC of indigenous peoples. Nagoya Protocol on Access to Genetic Resources and the Fair and Equitable Sharing of Benefits Arising from their Utilization to the Convention on Biological Diversity, Nagoya, 29 October 2010.

[21] See Permanent Forum on Indigenous Issues, Sixth session, New York, 14–25 May 2007, Item 4 of the provisional agenda, Report of the international expert group meeting on the international regime on access and benefit-sharing and indigenous peoples' human rights of the Convention on Biological Diversity, E/C.19/2007/8.

[22] Article 15.2.

[23] *Ibid.*

1989–1992, states, as well as many NGOs, echoed indigenous peoples' focus on land and other forms of tangible and intangible property as collective human rights, together with the right to self-determination.[24] Indigenous peoples' insistence on recognition of community ownership and control of land evolved into a more general principle: indigenous peoples have a major role to play in achieving sustainability as partners in decision-making and implementation on the ground. It follows that governments should respect the distinctive interests and perspectives of indigenous peoples and local communities, pursuing development through a more decentralized system of "partnership" with them.[25]

The Rio Conference has had a considerable impact on the development of indigenous peoples' human rights with a particular emphasis on land rights and participation. It can be convincingly argued that, after Rio, environmental issues have become a venue for indigenous peoples to strengthen their rights.[26] Important concepts, such as FPIC originate in the claim that indigenous peoples must retain the possibility to control the development of their lands and natural resources, and that this control leads to sustainable outcomes benefitting the larger environmental protection agenda.

The United Nations proclaimed 1993 the International Year of the World's Indigenous People, with a view to strengthening international cooperation in finding a solution to the problems faced by indigenous communities in areas such as human rights and the environment.[27] When the matter came before the Commission on Human Rights in the 1989–1990s session, the issue of whether to use the term "peoples" dominated the negotiations. Fearing that the use "peoples" in the title of the Year would imply the right to self-determination, Canada and Brazil insisted on the singular form. Consensus on the Year became contingent on accepting the position of these two countries.[28] The theme of the Year, "A New Partnership", also reflected a political compromise on the right to self-determination. Indigenous organizations had pressed for a reference to "self-determination" in the theme, but this was rejected by all of the governments as too provocative. Indigenous peoples, in turn, rejected themes based on "cultural diversity" as too weak.[29] This language battle shows the persistence of indigenous peoples to strengthen their status, which has turned to be successful tactic in international arenas.

---

[24] Barsh, R. L. (2007). Indigenous Peoples. In Bodansky, D. Brunnée, J. and Hey, E. (Eds.), The Oxford Handbook of International Environmental Law (pp. 830–831, at 839). Oxford University Press.

[25] *Ibid.*

[26] See Heinämäki, L. (2011). Towards an Equal Partnership between Indigenous Peoples and States: Learning from Arctic Experiences? The Yearbook of Polar Law, 3193–246.

[27] UNGA Resolution 45/164 of 18 December 1990. See also International Year of the World's Indigenous Peoples, 1993, UNGA Res. 48/133, 48 U.N. GAOR Supp. (No. 49) at 251, U.N. Doc. A/48/49 (1993).

[28] Barsh, R. L. (2004), *supra* note 16, at 25.

[29] *Ibid.*

Following the recommendation of the World Conference on Human Rights in December 1993, the General Assembly proclaimed the International Decade of the World's Indigenous People (1994–2004).[30] The General Assembly adopted an ambitious programme of activities and identified a number of specific objectives for the Decade: first and foremost the establishment of a permanent forum on indigenous issues and the adoption of the United Nations Declaration on the Rights of Indigenous Peoples.[31]

In December 2004, the Assembly renewed its commitment to promote and protect the rights of indigenous peoples, proclaiming the Second International Decade of the World's Indigenous People (2005–2014).[32] The emphasis of the new decade was to strengthen the active role of indigenous peoples in decision-making. The Decade promoted the idea of equal partnership. The main objectives set for the new Decade were to promote non-discrimination and the inclusion of indigenous peoples in all phases of the policy process and to promote a full and effective participation of indigenous peoples in decisions that affect their lives, based on the principle of FPIC.[33] These attempts had an important role to play in shifting the status of indigenous peoples towards legal personality.

The most important development connected with the International Decade was the establishment of the Permanent Forum on Indigenous Issues within the UN Economic and Social Council, which met for the first time in May 2002.[34] The establishment of the new Forum marked a fundamental milestone in the indigenous struggle to gain a position within the international community. The new body was unique in several ways, perhaps most importantly in the parity of its composition. The Forum is made up of 16 experts, 8 nominated by governments and the other 8 by indigenous organizations.[35] The Permanent Forum indicates the strengthened status of indigenous peoples in international law. The Forum operates at the highest possible level within the UN system. Its mandate is very broad: in fact, all the mandate areas of ECOSOC itself. It has been estimated that the Forum will provide for a previously lacking holistic approach to indigenous issues in the UN system, while it seeks to guarantee that all UN bodies, in all their activities, take the particular needs and concerns of indigenous peoples into account.[36] The UN Permanent Forum

---

[30] UNGA Res. 48/163 (Dec. 21, 1993) proclaiming the "International Decade of the World's Indigenous People", commencing Dec. 10, 1994.

[31] United Nations Economic and Social Council, Permanent Forum on Indigenous Issues, Seventh session, New York, 21 April-2 May 2008, Item 5 of the provisional agenda, Human rights: dialogue with the Special Rapporteur on the situation of human rights and fundamental freedoms of indigenous people and other special rapporteurs, E/C.19/2008/2, Para 15.

[32] UNGA Res. 59/174.

[33] *Ibid.*, Para. 16.

[34] ECOSOC Res. E/RES/2000/22 (July 28, 2000) establishing the Permanent Forum; Report of the First Session of the Permanent Forum on Indigenous Issues, U.N. Doc. E/2002/42/Supp. 43 (Wilton Littlechild, Rapporteur).

[35] *Ibid.*, Para. 25.

[36] Magga, O.H. (September 1, 2003). Presentation by the Chairperson of the Permanent Forum on Indigenous Issues, The UN Permanent Forum on Indigenous Issues – Ambitions and Limitations,

on Indigenous Issues at its first, second and third sessions, identified as a major methodological challenge the application of the principle of FPIC concerning indigenous peoples. At its third session, the body recommended the holding of a workshop on FPIC which was authorized by the ECOSOC.[37]

One of the key targets of the Permanent Forum on Indigenous Issues has been the push for an adoption of the UN Declaration on the Rights of Indigenous Peoples. The UN Declaration was adopted by the UN Human Rights Council on 29 June 2006 and by the UN General Assembly on 13 September 2007.[38] The UN Declaration indicates a historical shift in relation to the legal status of indigenous peoples in international law. For the first time, indigenous peoples participated in the drafting process of the actual text with an equal voice with governments.[39] Although the final decision-making was carried out in line with the general practice of international law, recognizing only states as parties to the instrument, it can be said that indigenous peoples, for the very first time in a global context, participated in the making of international law.

The UN Declaration is a clear step forward in the recognition of the rights and legal status of indigenous peoples in international law. In addition to the principles of self-development and cultural integrity adopted in ILO Convention No. 169, the Declaration celebrates a paradigm shift: not only does the Declaration explicitly recognise the right to self-determination and self-governance of indigenous peoples, but it also advances the right of FPIC in relation to decision-making concerning natural resources and other crucial matters. It was precisely these established rights that led the few key countries – Australia, Canada, New Zealand and the United States – to first vote against the Declaration in the General Assembly.[40] Later on, however, all these countries have endorsed the Declaration, although some of them are still reluctant to fully embrace the right to FPIC. Although the UN Declaration is not a legally binding instrument, human rights monitoring bodies

---

Resource Centre for the Rights of Indigenous Peoples, Guovdageaidnu. Retrieved January, 19, 2014, from http://www.galdu.org/web/index.php?artihkkal=39&giella1=eng

[37] International Workshop on free, prior and informed consent and indigenous peoples (New York, 17–19 January 2005), PFII/2005/WS. 2/3: Contribution of the Convention on Biological Diversity and the Principle of Prior and Informed Consent, Secretariat of the Convention on Biological Diversity.

[38] The United Nations Declaration on the Rights of Indigenous Peoples, 7 September 2007, Sixty-first Session, A/61/L.67.

[39] See Davis, M. (2008). Indigenous Struggle in Standard-Setting: The United Nations Declaration on the Rights of Indigenous Peoples. *Melbourne* Journal of International Law, 9(2), 1–33, at 2. See also, generally, Barsh, R.L. (1996). Indigenous Peoples and the UN Commission on Human Rights: A Case of the Immovable Object and the Irresistible Force. 18 Human Rights Quarterly, 782–813.

[40] Tauli-Corpuz, V. (2007). The Declaration on the Rights of Indigenous Peoples: A major victory and a challenge. Retrieved January 19, 2014, from http://www.twnside.org.sg/title2/resurgence/206/cover1.doc

have already started to apply it as a legal source, as will be discussed in this article.[41]

Today, indigenous peoples are working towards the implementation of the Declaration. Work is currently being done with the UN Permanent Forum on Indigenous Issues to determine how the Declaration should be implemented by the UN and its agencies, and how the Declaration's implementation by Member States can be assessed.[42] This is done together with the relatively new Expert Mechanism on the Rights of Indigenous Peoples (EMRIP) that was created in 2007 to replace and continue the work of the WGIP.[43]

The right to self-determination, as understood in the UN Declaration, does not seem to give a total freedom to determine indigenous peoples' political status, while strongly protecting the integrity of sovereign states.[44] Fitzmaurice rightfully states that "the definition of self-determination in the Declaration is considered to be a compromise between the aspirations of indigenous peoples and the reluctance of States to grant a broadly understood right to self-determination".[45] Thus, according to the UN Declaration, self-determination does not entail the right to secession. On the other hand, an argument can be made that since the Declaration now recognizes indigenous peoples as "peoples", they should, accordingly, enjoy the rights of peoples under international law.[46] In relation to this view, an argument has been raised

---

[41] The Supreme Court of Belize also applied the principles of the Declaration as a framework for determining land rights. Shortly after the adoption of the Declaration by the UN General Assembly, the Supreme Court of Belize made a decision relating to the rights of the Maya community to their lands and resources, applying the Declaration. *Aurelio Cal v. Attorney-General of Belize Claim* 121/2007, 18 October 2007, Supreme Court of Belize, http://www.elaw.org/node/1620 (accessed 19 January 2014). Finally, it should be mentioned that Bolivia was the first country to adopt the UN Declaration on the Rights of Indigenous Peoples as national law. National Law 3760, which is an exact copy of the UN Declaration, was passed on November 7, 2007. See IWGIA, http://www.iwgia.org/sw18043.as (accessed 19 January 2014).

[42] See International expert group meeting on the role of the United Nations Permanent Forum on Indigenous Issues in the implementation of Article 42 of the United Nations Declaration on the Rights of Indigenous Peoples, 14–16 January, New York, United Nations Department of Economic and Social Affairs, Division for Social Policy and Development, Secretariat of the Permanent Forum on Indigenous Issues, PFII/2009/EGM1/15, http://www.un.org/esa/socdev/unpfii/documents/EGM_Art_42_FAO.doc (accessed 19 January 2014).

[43] Human Rights Council Resolution 6/36. Expert Mechanism on the Rights of Indigenous Peoples, 6th Session, 14/12/2007, A/HRC/RES/6/36. The mandate of the EM is to provide its thematic expertise in the manner and form requested by the Human Rights Council. It will focus mainly on studies and research-based advice.

[44] As stated by Daes, "The principle of self-determination as discussed within the Working Group and as reflected in the draft declaration was used in its internal character, that is short of any implications which might encourage the formation of independent States." See UN Doc. E/CN.4/Sub2AC.4/1992/3 Add. 1, at 5 (1992). See Article 46 of the UN Declaration on the Rights of Indigenous Peoples, which explicitly protects the territorial integrity of states.

[45] See Fitzmaurice, M. (2009). The New Developments Regarding the Saami Peoples of the North. Journal on Minority and Group Rights, 16, 67–156, at 151.

[46] See Koivurova, T. (2008). Alkuperäiskansojenitsemääräämisoikeuskansainvälisessäoikeudessa [The right of self-determination of indigenous peoples in international law]. In Aarto M. $&

that although in non-colonial territories the right to self-determination does not amount to a right for a part of the population to secede from existing states, there might be exceptional circumstances in which a group may have a legally and politically tenable right to secession due to their demonstrable inability to achieve the established right of self-determination guaranteed by law.[47]

What the UN Declaration does, however, recognize, is the full self-determination as far as the economic, social and cultural development of indigenous peoples is concerned. The Declaration guarantees the right to self-government in internal and local matters.[48] In addition, effective and meaningful participation – the right to consultation or even FPIC with respect to land and resource use and other important matters, such as participation in international decision-making – has a key role to play in the determination of economic, social and cultural development. The right to FPIC been seen as a part of the "new" self-determination of indigenous peoples, as will be discussed later.

It can convincingly be argued that a paradigm shift has emerged in relation to the international status of indigenous peoples, which is currently further evolving. From the original viewpoint of seeing indigenous peoples as passive objectives of protection, mainly seeking to guarantee their equal enjoyment of human rights, and from a cautious recognition of their unique culture, many important steps have been taken towards guaranteeing their authoritative position and effective participation in the matters that are important to them. Clear shift has taken place from minority protection and general non-discrimination towards the recognition of indigenous communities as peoples. During the past years, indigenous peoples have been guaranteed "*sui generis*" rights, namely because of their particular nature-related way of life. Both human rights monitoring bodies and biodiversity protection regimes recognize this special status for indigenous peoples due to their connection to nature and natural resources. Specifically, this has happened in relation to the protection of the lands of indigenous peoples, which has led to the recognition of fundamental rights, such as rights to self-determination and FPIC. International standards are celebrating a beginning of a new era in state-indigenous relations, which still has to be implemented in concrete ways in the domestic and local settings.

3 Concept of Free, Prior and Informed Consent (FPIC) of Indigenous Peoples

The right to free, prior and informed consent (FPIC) in relation to resource extraction and other development projects within the territories traditionally occupied and used by indigenous peoples is currently a very topical issue internationally,

Vartiainen, M. (Eds), Oikeus kansainvälisessä maailmassa [Law in a changing world] (pp. 249–269, at 268). Edita Publishing Oy, Lapin yliopiston oikeustieteiden tiedekunta [University of Lapland, Faculty of Law].

[47] See *Reference re Secession of Quebec* (1998) 2 S.C.R. 217.

[48] Article 4.

regionally and domestically. As maintained by the study of the Commission on Human Rights, discussion and standard settings over this topic cover a wide range of bodies and sectors ranging from the safeguard policies of the multilateral development banks and international financial institutions; practices of extractive industries; water and energy development; natural resources management; access to genetic resources and associated traditional knowledge and benefit-sharing arrangements; scientific and medical research; and indigenous cultural heritage.[49]

On a basic level, the concept of FPIC can be translated as the right of indigenous peoples to make free and informed choices about the development of their lands and resources.[50] The basic idea of FPIC is to make sure that indigenous peoples are not forced or threatened, and that their consent is asked and freely given prior to the authorization or beginning of any activities that take place in their traditional lands and which could have negative impacts on them. Ultimately, the principle of FPIC signifies that the choices of indigenous peoples to give or withhold consent are respected.[51]

Today, indigenous peoples in many places of the world are trying to renegotiate their relations with states and with new private sector operations seeking access to the resources on their lands.[52] They are asserting their right to FPIC as exercised through their representative institutions in dealing with the many parties interested in their traditional territories. Indigenous peoples are seeking support from international human rights bodies to find new ways of being recognized by international and national laws and systems of decision-making with the recognition of their autonomy and their own values.[53]

In relation to development projects affecting indigenous peoples' lands and natural resources, the respect for the principle of FPIC is important so that: (1) indige-

---

[49] U.N. Commission on Human rights, Sub-Comm. On the Promotion and Protection of Human Rights Working Group on Indigenous Populations, Working Paper: Standard-Setting: Legal Commentary on the Concept of Free, Prior and Informed Consent, 57, U.N. Doc. E/CN.4/Sub.2/AC.4/2005/WP.1, 2005 (July 14, 2005), at 3. (prepared by Antoanella-Iulia Motoc and the Tebtebba Foundation).

[50] Ward, T. (2011). The Right to Free, Prior and Informed Consent: Indigenous Peoples' Participation Rights within International Law. Northwestern Journal of International Human Rights, 10(2), 54.

[51] U.N. Commission on Human rights, Sub-Comm. On the Promotion and Protection of Human Rights Working Group on Indigenous Populations, Working Paper: Standard-Setting: Legal Commentary on the Concept of Free, Prior and Informed Consent, 57, U.N. Doc. E/CN.4/Sub.2/AC.4/2005/WP.1, 2005 (July 14, 2005) (prepared by Antoanella-Iulia Motoc and the Tebtebba Foundation).

[52] U.N. Commission on Human rights, Sub-Comm. On the Promotion and Protection of Human Rights Working Group on Indigenous Populations, Working Paper: Standard-Setting: Legal Commentary on the Concept of Free, Prior and Informed Consent, 57, U.N. Doc. E/CN.4/Sub.2/AC.4/2005/WP.1, 2005 (July 14, 2005), at 4.

[53] See Marcus Colchester & Mackay, F. (2004). In Search of Middle Ground: Indigenous Peoples, Collective Representation and the Right to Free, Prior and Informed Consent. Forest Peoples Program. Retrieved January 19, 2014, from http://www.forestpeoples.org/sites/fpp/files/publication/2010/08/fpicipsaug04eng.pdf

nous peoples are not coerced, pressured or intimidated in their choices of development; (2) their consent is sought and freely given prior to the authorization and start of development activities; (3) indigenous peoples have full information about the scope and impacts of the proposed development activities on their lands, resources and well-being; (4) their choice to give or withhold consent over developments affecting them is respected and upheld.[54]

From a legal or technical perspective, however, FPIC is a much contested and confused concept. There are both non-binding and binding international legal instruments and industry standards that purport to require some form of FPIC. As a result, its definition, including terms such as "land", "territories", "significant impact", are subjects to numerous conflicting interpretations and requirements, depending "which FPIC" is at issue.[55]

Especially after the adoption of the UN Declaration on the Rights of Indigenous Peoples, the right to FPIC has been directly related to the right of indigenous peoples to self-determination. As maintained by Ward, FPIC and other participation rights are not merely administrative processes, but are an exercise in and expression of the right to self-determination.[56] For indigenous rights advocates, FPIC and other participatory rights are derived from the right to self-determination, which is considered to be the fundamental principle of indigenous peoples' rights.[57] Ward, for instance, sees that FPIC within the UN Declaration is conceived of as a way to ensure that the right to self-determination is respected and protected by states.[58]

As argued by the report of the UN Commission on Human Rights, self-determination of peoples and the corollary right of FPIC, is integral to indigenous peoples' control over their lands and territories, to the enjoyment and practice of their cultures, and to make choices over their own economic, cultural and social development. This right, in order to be meaningful, must include the right to withhold consent to certain development projects or proposals. Self-determination and FPIC, as collective rights, fundamentally entail the exercise of choices by peoples, as rights-bearers and legal persons about their economic, social and cultural development. These cannot be weakened to consultation of individual constituents about their wishes, but rather must enable and guarantee the collective decision-making of

[54] U.N. Commission on Human rights, Sub-Comm. On the Promotion and Protection of Human Rights Working Group on Indigenous Populations, Working Paper: Standard-Setting: Legal Commentary on the Concept of Free, Prior and Informed Consent, 57, U.N. Doc. E/CN.4/Sub.2/AC.4/2005/WP.1, 2005 (July 14, 2005), at 3. (prepared by Antoanella-Iulia Motoc and the Tebtebba Foundation), at 15, Para 57.

[55] Seier, F. (2011). Free, Prior and Informed Consent' under UNDRIP: What Does it Really Mean? Right 2 Respect, Business and Human Rights Advisors. Retrieved January 19, 2014, from http://www. right2respect. com/2011/06/%E2%80%98free-prior-and-informed- consent%E2%80%99-under-the-un-declaration-on-the-rights-of-indigenous-peoples-what-does-it-really-mean

[56] Ward, T. (2011), *supra* note 50, at 55.

[57] Bartolome, C. (2005) The Indigenous Rights of Participation and International Development Policies. Arizona Journal of International & Comparative Law, 41, at 41; U.N. Commission on Human Rights (2005), supra note 56, at 56.

[58] Ward, T. (2011), *supra* note 50, at 58.

the concerned indigenous peoples and their communities through legitimate customary and agreed processes, and through their own institutions.[59]

While for indigenous rights advocates self-determination is the basis for FPIC, within international human rights jurisprudence FPIC is legally based in property rights, cultural rights, and the right to non-discrimination.[60] These rights, although recognizing a collective element in the case of indigenous peoples, have an individual rather than collective basis. UN Human Rights Committee (HRC), for instance, accepts communications from individuals concerning individual human rights. According to the case practice of the HRC, it receives only complaints based on individual rights, such as a right of members of a minority group in Article 27, but not a right of self-determination (Art. 1) that is a right of a collective.[61] FPIC has become acknowledged recently as a part of Article 27, as will be discussed. Additionally, FPIC, as adopted as a part of the biodiversity regimes, is not directly connected to the question of self-determination, but to the acknowledgment that indigenous peoples, as holders of traditional knowledge, can make a valuable contribution to the biodiversity protection, and thus should participate and share benefits of the use genetic resources. Therefore, it can still be debated whether FPIC should be linked only to the question of self-determination, or whether it is more meaningful to talk about it also as an inherent part of the right to cultural integrity and the right to property.

3.1 FPIC in Indigenous Rights Instruments

The ILO Convention No. 169 refers to the right of FPIC in the context of relocation of indigenous peoples from their land in its article 16. Article 7 recognize indigenous peoples' "right to decide their own priorities for the process of development" and "to exercise control, to the extent possible, over their own economic, social and cultural development." In articles 2, 5 and 15, the Convention requires that States fully consult with indigenous peoples and ensure their informed participation in the context of development, national institutions and programs, and the management of lands and resources. As a general principle, Article 6 requires that consultation must be undertaken in good faith, in a form appropriate to the circumstances and with the objective of achieving consent. It is thus argued that Articles 6 and 7 of the ILO

[59] U.N. Commission on Human rights, Sub-Comm. On the Promotion and Protection of Human Rights Working Group on Indigenous Populations, Working Paper: Standard-Setting: Legal Commentary on the Concept of Free, Prior and Informed Consent, 57, U.N. Doc. E/CN.4/Sub.2/AC.4/2005/WP.1, 2005 (July 14, 2005) (prepared by Antoanella-Iulia Motoc and the Tebtebba Foundation), at 12, Para 45.

[60] Ward, T. (2011), *supra* note 50, at 56.

[61] See, *Lubicon Lake Band v. Canada*, Communication No. 167/1984, CCPR/C/38/D/167/1884.

Convention "reflect the spirit of prior informed consent and apply to each provision of ILO 169".[62]

In applying the Convention, the ILO has held that consultations must be held when a variety of indigenous interests are involved, including legislative measures regulating the consultation process itself, constitutional provisions concerning indigenous peoples, development of lands adjacent to, or in indigenous territories, to the complete destruction of those lands.[63]

Participation rights have been the foundation of the ILO Committee of Experts, CEACR's[64] interpretations of how the Convention applies to State Parties.[65] For example, CEACR, while examining Ecuador's non-compliance with the Convention stated that "the spirit of consultation and participation constitutes the cornerstone of ILO Convention No. 169 on which all its provisions are based".[66] Cases concerning oil exploration concessions in Ecuador, the ILO Committee emphasized article 6 (2), which requires that consultations must be in good faith, through culturally appropriate procedures, and with the objective of reaching an agreement with the affected indigenous peoples. The CEACR stated: "The concept of consulting the indigenous communities that could be affected by the exploration or exploitation of natural resources includes establishing a genuine dialogue between both parties characterized by communication and understanding, mutual respect, good faith and the sincere wish to reach a common accord".[67] The CEACR has repeatedly called on State Parties to respect their obligations to consult with indigenous peoples prior to exploration and exploitation of natural resources within their traditional territories,

---

[62] Baluarte, D.C. (2004). Balancing Indigenous Rights and a State's Right to Develop in Latin America: The Inter-American Rights Regime and ILO Convention 169, 4 Sustainable Development Law & Policy, 9, at 10.

[63] See Anaya (2005), *supra* note 17, at 11. See Report of the Committee Set up to Examine the Representation Alleging Non-Observance by Colombia of the Indigenous and Tribal Peoples Convention, 1989 (No. 169), Made Under Article 24 of the ILO Constitution by the Central Unitary Workers' Union (CUT), ILO Doc. GB. 282/14/2 (Nov. 21, 2001); Report of the Committee Set Up to Examine the Representation Alleging Non-Observance by Mexico of the Indigenous and Tribal Peoples Convention, 1989 (No. 169), Made Under Article 24 of the ILO Constitution by the Union of Workers of the Autonomous University of Mexico (STUNAM) and the Independent Union of Workers of La Jornada (SITRAJOR). ILO Doc. GB.289/17/3 (Mar. 19, 2004); Report of the Committee Set Up to Examine the Representation Alleging Non-Observance by Colombia of the Indigenous and Tribal Peoples Convention, 1989 (No. 169), Made Under Article 24 of the ILO Constitution by the Central Unitary Workers' Union (CUT) and the Colombian Medical Trade Union Association, ILO Doc. GB.282/14/3 (Nov. 14, 2001).

[64] The Committee of Experts on the Application of Conventions and Recommendations.

[65] Ward, T. (2011), *supra* note 50, at 60.

[66] ILO, Report of the Committee set up to examine the representation alleging non-observance by Ecuador of the Indigenous and Tribal Peoples Convention, 1989 (No. 169), made under article 24 of the ILO Constitution by the ConfederacionEcuatoriana de OrganizacionesSindicalesLibres (CEOSL), ILO Doc. GB.282/14/2 (Nov. 14, 2001).

[67] Report of the Committee Set Up to Examine the Representation Alleging Non-Observance by Ecuador of the Indigenous and Tribal Peoples Convention, 1989 (No. 169), Made Under Article 24 of the ILO Constitution by the ConfederacionEcuatoriana de OrganizacionesSindicalesLibres (CEOSL), ILO Doc. GB.282/14/2 (Nov. 14, 2001), Para 38.

and has required on the adoption and implementation of domestic legislation in order to facilitate such consultations.[68]

The UN Declaration explicitly calls for the FPIC of indigenous peoples in: Article 10 in the case of relocation of indigenous communities, Article 19 when a State is adopting legislative or administrative measures that affect indigenous peoples, and Article 29 regarding the disposal of hazardous waste within their territories. In addition, Article 32 requires free and informed consent prior to "the approval of any project affecting their lands or territories and other resources, particularly in connection with the development, utilization or exploitation of mineral, water or other resources."

If one compares the language of Articles 19 and 32, on the one hand, with Articles 10 and 29, on the other, an argument could be made that the first two articles simply contemplate a good faith consultative and cooperative process "in an effort to, but not necessarily" obtain the indigenous peoples' consent, while the latter two articles, which don't articulate any such process, provide for an absolute prohibition on certain activities "unless FPIC has been obtained" (i.e. *veto*).[69]

During the lengthy negotiations of the UN Declaration, participation rights were some of the most contentious in large part because of the ambiguity of the definition.[70] For some indigenous rights advocates FPIC is seen as a right to *veto* projects, while others argue that FPIC is not meant to be a *veto* right, but rather a way of ensuring that indigenous peoples meaningfully participate in decisions directly impacting their lands, territories and resources.[71]

While analyzing the provisions concerning FPIC in the UNDRIP, the Expert Mechanism on the Rights of Indigenous Peoples has distinguished the requirement of FPIC into mandatory and contextual requirement. Mandatory requirement is set in Article 10 of the Declaration that prohibits the forcible removal of indigenous peoples from their lands. Additionally, article 29 states that "states shall take effective measures to ensure that no storage or disposal or hazardous materials shall take place in the lands or territories of indigenous peoples without their free, prior and informed consent." In other cases, according to the Expert Mechanism, the requirement to obtain FPIC depends on context, including, notably, in relation to the approval of projects affecting indigenous peoples' lands, territories and other resources, referring to Article 32 of UNDRIP. In the final report on its study on

---

[68] See CEACR, Individual Observations concerning Indigenous and Tribal Peoples Convention, 1989 (No. 169) Ecuador, ILO Doc. 062010ECU 169 (2010), para 4; CEACR, Individual Observation concerning Indigenous and Tribal Peoples Convention, 1989 (No. 169) Guatemala, ILO Doc. 062006GTM169 (2006), Paras 10,13 and 15; CEACR, Individual Observation concerning Indigenous and Tribal Peoples Convention, 1989 (No. 169) Mexico, ILO Doc. 062006MEX169 (2006), Para 10.

[69] Seier, F. (2011), *supra* note 55, at 2.

[70] Davis, M. (2008), *supra* note 39, at 465.

[71] Special Rapporteur James Anaya on the Rights of Indigenous Peoples, Promotion and Protection of All Human Rights, Civil, Political, Economic, Social and Cultural Rights, Including the Right to Development: Report of the Special Rapporteur on the Situation of Human Rights and Fundamental Freedoms of Indigenous People, UN Doc. A/HRC/13/34 (July 15, 2009), Para 48.

indigenous peoples and the right to participate in decision-making, the Expert Mechanism provides further clarification:

The Declaration on the Rights of Indigenous Peoples requires that the free, prior and informed consent of indigenous peoples be obtained in matters of fundamental importance to their rights, survival, dignity and well-being. In assessing whether a matter is of importance to the indigenous peoples concerned, relevant factors include the perspective and priorities of the indigenous peoples concerned, the nature of the matter or proposed activity and its potential impact on the indigenous peoples concerned, taking into account, inter alia, the cumulative effects of previous encroachments or activities and historical inequities faced by the indigenous peoples concerned.[72]

The analysis of the Expert Mechanism is supported by both Inter-American Court of Human Rights as well as the UN Human Rights Committee, of which both have recognized that in the case of significant, large-scale negative impact on the lands and traditional way of life of indigenous peoples, mere consultation is not enough but the FPIC has to be gained.[73] Therefore, it can be argued that while ILO Convention No. 169 requires "the spirit of FPIC", UN Declaration does in fact require the "body of FPIC", which means that the principle of FPIC is applied in the cases where interference would cause a significant negative impact on the traditional lands and way of life of indigenous peoples.

Although the Declaration is not a legally binding document, it has been argued that it already affirms existing customary international law.[74] Others argue, quite rightfully according to the present author, that it is not completely accurate to suggest that the Declaration already represents customary international law.[75] The idea behind a declaration and other non-binding instruments is that they create norms that can guide the behavior of states and ultimately this behavior may develop into customary international law.[76]

What the Declaration affirms in practice are the present views of the states in relation to many important rights of indigenous peoples. As stated by Fitzmaurice, the Declaration is a long-awaited affirmation of the position of indigenous peoples as important actors in the contemporary world whose interests must be taken into account by states. Additionally, it is a further indication that states and indigenous peoples must be engaged in a dialogue.[77]

---

[72] A/HRC/18/42, Para 22.

[73] See the next chapter.

[74] Bartolome Clavero (2001), *supra* note 57, at 43.

[75] Davis, M. (2008), supra note 39, at 465; Xanthaki, A. (2009). Indigenous Rights in International Law over the Last 10 Years and Future Developments, 10 Melbourne Journal of International Law, at 36.

[76] Anaya, J. (2009). The Right of Indigenous Peoples to Self-Determination in the Post-Declaration Era. In Chartres C. & Stavanhagen, T. (Eds.), Making the Declaration Work: The United Nations Declaration on the Rights of Indigenous Peoples (at 184), IWGIA.

[77] Fitzmaurice, M. (2009), *supra* note 45, at 76.

Despite the positive steps of the countries that originally rejected the Declaration, the later endorsement of this instrument by some of the countries or companies or institutions operating within those countries still appears rather hesitant. At the time of its endorsement of the UN Declaration in December 2010, the United States indicated that FPIC calls for "a process of meaningful consultation with tribal leaders, but not necessarily … agreement …, before the actions addressed in those consultations are taken".[78]

In March 2011, the government of Canada issued its "Updated Guidelines for Federal Officials to Fulfill the Duty to Consult". These Guidelines fail to consider the right of Indigenous peoples to FPIC, except to indicate Canada's concern when such consent is "interpreted as a veto".[79] In May the same year, at the Commission on Sustainable Development's Working Group on Mining, Canada, Australia, New Zealand and the United States asked for deletion of "free, prior and informed consent" regarding indigenous and local communities.[80] On the other hand, these kinds of statements should kept in mind that the Canadian Supreme Court, in *Delgamuukw v. British Colombia*[81] has argued that, where indigenous property rights are proven, the duty to consult varies with the circumstances, from a duty to discuss important decisions where proposed breach is relatively minor, to full consent for very serious issues.

UN Permanent Forum on Indigenous Issues reminds that in Canada, a number of large banks have recently trumpeted their adoption of FPIC, when what they really mean is free, prior and informed consultation.[82] The Permanent Forum expresses its justifiable concern that "momentum toward implementation of free, prior and informed consent may be lost or diverted by the misleading use of the inadequate standard of free, prior and informed consultation".[83] The concern of the Permanent Forum seems a very real one: if the requirement of the consent is replaced with the duty of the consultation, the UN Declaration would really not have established anything more than the ILO Convention No 169 in terms of this crucial matter. As the Permanent Forum points out: "with the lesser standard of consultation would mean that at the conclusion of such a process taking place, governments or corporations would continue to be free to act in their own interests and the interests of other powerful sectors of society – while unilaterally and arbitrarily ignoring the decision taken by Indigenous peoples. This is contrary to the very purpose of FPIC".[84]

[78] See Permanent Forum on Indigenous Issues, Joint Statement: Free, prior and informed consent, Tenth Session, May 18, 2011, Agenda Item 3 c: Free, prior and informed consent, Speaker: Kenneth Deer, available at: http://www.ubcic.bc.ca/News_Releases/UBCICNews05191101.html#ixzz21qDLwJWr (visited 23.10.2012).

[79] *Ibid.*

[80] *Ibid.*

[81] 153 D.L.R.4th 193 (1997).

[82] *Ibid.*

[83] *Ibid.*

[84] *Ibid.*

Although any profound analysis of the domestic legal sources is beyond the scope of this article, it should be shortly mentioned that several domestic legal instruments underline the importance of FPIC. For example, in five states of Australia, consent must be obtained in connection with mining through statutory indigenous-controlled Land Councils for more than 30 years.[85] Additionally, Aotearoa-New Zealand Crown Minerals Act[86] guarantees a special protection for Maori land, as defined by the TeTureWhenua Maori Act 1993: if the Maori land is regarded as waahitapu (sacred areas), access even for minimum impact activities can only be obtained if the Maori landowners give their consent.[87] For activities other than minimum impact activities, the owners of Maori land also have a right to consent[88] even where there may be public interest grounds that would require arbitration in the case of non-Maori land owners.[89]

4 FPIC and Human Rights Monitoring Bodies

4.1 General Comments and Concluding Observations

The Committee on Economic, Social and Cultural Rights (CESCR) recognizes indigenous peoples' collective rights to lands and resources through their right to participate in and maintain their cultures.[90] CESCR has, on a number of occasions, highlighted the need to obtain indigenous peoples' consent in relation to resource exploitation in their traditional lands. In 2004, the Committee stated that it was "deeply concerned that natural extracting concessions have been granted to international companies without the full consent of the concerned communities".[91] In 2007, The Committee has also observed "with regret that traditional lands of indigenous peoples have been reduced or occupied, *without their consent*, by timber, mining and oil companies, at the expense of the exercise of their culture and the equilibrium

[85] Aboriginal Land Rights (Northern Territory) Act 1976, Pt. IV; Aboriginal Land Rights Act 1983 (NSW), sec. 45(5); Aboriginal Land Act 1991 (Qld), sec. 42; and Torres Strait Islander Land Act 1991 (Qld), sec. 80; Mineral Resources Act 1989 (Qld), sec. 54; Mineral Resources Development Act 1995 (Tas), Pt. 7, and; Aboriginal Land (Jervis Bay Territory) Act 1986 (Cth), sec. 43, 52 A (1), (2).

[86] Aotearoa-New Zealand's, Crown Minerals Act 1991.

[87] Section 51.

[88] Sections 53–54.

[89] The Philippine Indigenous Peoples Rights Act recognizes the right of free, prior and informed consent of indigenous peoples for all activities affecting their lands and territories including: exploration, development and use of natural resources; research and bio-prospecting; displacement and relocation; archaeological explorations; policies affecting indigenous peoples such as Executive Order 263 (Community-based Forest Management); and the entry of military.

[90] Ward, T. (2011), *supra* note 50, at 57.

[91] CESCR, Concluding observations of the Committee on Economic, Social and Cultural Rights: Ecuador, U.N. Doc. E/C.12/1/Add. 100, para 12.

of the ecosystem (emphasis added)".[92] It subsequently recommended that the State Party ensure the participation of indigenous peoples in decisions affecting their lives and particularly urged it to consult and seek the consent of the indigenous peoples concerned prior to the implementation of timber, soil and subsoil mining projects and on any public policy affecting them.[93]

After the adoption of the UN Declaration on the Rights of Indigenous Peoples, the CESCR has further expanded on the right of indigenous peoples to FPIC in General Comment No. 21.[94] This interpretation of Article 15 of International Covenant on Economic, Social and Cultural Rights (ICESCR), which outlines the right to participate in cultural life, includes the rights of indigenous peoples to restitution or return of lands, territories and resources traditionally used and enjoyed by indigenous communities if taken "without the prior and informed consent of the affected peoples."[95] Furthermore, it calls on State Parties to the Convention to "respect the principle of free, prior and informed consent of indigenous peoples in all matters covered by their specific rights."[96]

The Committee on the Elimination of Racial Discrimination (CERD) has also recognized the requirement for the consent of indigenous peoples. In its General Recommendation 23, CERD called on states to "recognize and protect the rights of indigenous peoples to own, develop, control and use their communal lands, territories and resources,"[97] in fulfillment of the non-discrimination norm. CERD further requires states to "ensure that members of indigenous peoples have equal rights in respect of effective participation in public life and that no decisions directly relating to their rights and interests are taken without their informed consent."[98]

In its 2008 Concluding Observations on Russia, CERD recommended that the Government of Russia "seek the free informed consent of indigenous communities and give primary consideration to their special needs prior to granting licenses to private companies for economic activities on territories traditionally occupied or used by those communities."[99]

In 2012, CERD is again clear about the acknowledgement of the FPIC. In relation to Sámi indigenous people in Finland, it expresses a concern about the unresolved issue of the land rights of Sámi, which has not been satisfactorily settled, and

---

[92] CESCR, Concluding observations on Colombia, U.N. Doc. E/C.12/1/Add.74E/C.12/1/Add.74, para 12.

[93] *Ibid*, Para 33.

[94] CESCR, General Comment No. 21 Right of everyone to take part in cultural life (art. 15, 1(a), of the International Covenant on Economic, Social and Cultural Rights), U.N. Doc. E/C.12/GC/21 (Dec. 21, 2009).

[95] *Ibid*, Para 36.

[96] *Ibid*, Para 37.

[97] General Recommendation 23: Indigenous Peoples, Committee for the Elimination of Racial Discrimination, U.N. Doc. A/52/18, annex V; CERD/C/51/Misc.13/Rev.4 (1997), Para 5.

[98] *Ibid*, Para 4 (d).

[99] See UN Doc. CERD/C/RUS/CO/19, 20 August 2008 Concluding observations of the Committee on the Elimination of Racial Discrimination Russian Federation 73rd CERD session.

that various projects and activities, such as mining and logging, continue to be carried out in the traditional lands of Sámi people without their prior, free and informed consent.[100]

United Nations Human Rights Committee (HRC) seems to be under a process of shifting its approach from the original requirement of consultation towards the recognition of the FPIC. Although in its jurisprudence, FPIC has already been recognized by HRC in 2009, as will be described in the following chapter, in its concluding observations to the states' reports, the Committee is still somewhat inconsistent, although a trend seems to be an opening towards the recognition of FPIC.

In 2009, the HRC required only consultation from State parties in relation to the rights of indigenous peoples, for instance in its Concluding Observations on Australia[101] and Sweden.[102] A year after, however, in its Concluding Observation on El Salvador, the HRC, while urging the State Party to consider ratifying the ILO convention No. 169, states:

> Following consultations with all indigenous peoples, and with their free and informed consent, the State party should include in the next population census questions relating to the identification of indigenous peoples; design and implement public policies to move towards the full realization of their rights; and adopt special measures to address their marginalization. The State party should also, after consultation with all indigenous peoples, adopt measures to revive their languages and cultures.[103]

In 2011, the HRC noted "with concern that neither the existence of indigenous peoples in Togo nor their right to free, prior and informed consent is recognized." It stated furthermore that the State party should take the necessary steps to guarantee the recognition of minorities and indigenous peoples, and should ensure that indigenous peoples are able to exercise their right to FPIC.[104]

In 2012, the HRC sort of took a step back and adopted a somewhat confusing concept of "free and informed consultation" in its Concluding Observations on Guatemala. It stated:

[100] Committee on the Elimination of Racial Discrimination, Eighty-first session, 6–31 August 2012, Consideration of reports submitted by State parties under article 9 of the convention, Concluding observation of the Committee on the Elimination of Racial Discrimination, Finland, CERD/C/FIN/CO/20-22, Para 13. See also Report of the Committee on the Elimination of Racial Discrimination, U.N. GAOR, 50th Sess., Agenda Item 103, Para 536, U.N. Doc. A/50/18 (1995); See, also, Committee on the Elimination of Racial Discrimination, Concluding observations of the Committee on the Elimination of Racial Discrimination: Ecuador, U.N. Doc. CERD/C/ECU/CO/19, Para 16.

[101] UN Human Rights Committee, Concluding Observations on Australia, 7 May 2009, CCPR/C/AUS/CO/5.

[102] UN Human Rights Committee, Concluding Observations on Sweden, 7 May 2009, CCPR/C/SWE/CO/6.

[103] UN Human Rights Committee, Concluding observations on El Salvador, 18 November 2010, CCPR/C/SLV/CO/6 Para 18.

[104] UN Human Rights Committee, Concluding Observations on Togo, 18 April 2011, CCPR/C/TGO/CO/4, Para 21.

> The State party should comply with its international commitment to carry out prior and informed consultations with indigenous peoples for all decisions relating to projects that affect their rights, in accordance with article 27 of the Covenant. The State party should also recognize and take due account of all decisions taken by indigenous peoples during such consultations.[105]

In the Concluding Observations on Finland, 2013, the HRC has used another concept "free, prior and informed participation" when requesting the State of Finland to increase its efforts to revise its legislation to fully guarantee the rights of the Saami people in their traditional land.[106] It should be mentioned that Finland is currently in a process to strengthen the status of the Saami Parliament in several legislative acts.[107] Because the HRC is directly relating to this process, it may deliberately use a more cautious language, since it is quite clear that the State of Finland is still reluctant to recognize the consent of indigenous peoples in a way that indicates veto powers.

A different tone has been expressed by the HRC in relation to the Inter-American States where the concept of FPIC is more generally used, both by the human rights machinery as well as by national states. For instance, in concluding observations on Belize in 2013, referring to the decision of the Supreme Court of Belize, the HRC prohibits the State Party from issuing new concessions for logging parceling for private leasing, oil drilling, seismic surveys and road infrastructure projects in Mayan territories "without the free, prior, and informed consent of the relevant Mayan community."[108]

The Inter-American human rights Court and Commission have also extensively dealt with the issue of consultation and consent in both the Country Reports and their jurisprudence. As early as 1984, the Inter-American Commission on Human Rights stated, in relation to the Country Report concerning Nicaragua, that "the 'preponderant doctrine' holds that the principle of consent is of general application

[105] UN Human Rights Committee, Concluding observations on Guatemala, 19 April 2012, CCPR/C/GTM/CO/3, Para 27.

[106] UN Human Rights Committee, Concluding observations on Finland, 22 August 2013, CCPR/C/FIN/CO/6. Para 16.

[107] Finnish Mining Act 621/2011. See an analysis on the rights of the Sami people, Timo Koivurova and Anna Petrétei: Enacting a New Mining Act in Finland – How were Sami Rights and Interests Taken into Account? Nordisk Miljörättslig Tidskrift, Nordic Environmental Law Journal, 2014:1, 119–133, <http://nordiskmiljoratt.se/onewebmedia/NMT%202014-1.pdf> accessed 1 September 2014; Saamelaiskäräjälakityöryhmän mietintö. Oikeusministeriö, mietintöjä ja lausuntoja 55/2013. <http://oikeusministerio.fi/fi/index/julkaisut/julkaisuarkisto/1382513081296/Files/OMML_55_2013_MIETINTO_196_s.pdf> accessed in 20 August 2014; Reform in Act on Metsähallitus (State Forrest Board), Työryhmämuistio MMM 2014: 2. <http://www.mmm.fi/attachments/mmm/julkaisut/tyoryhmamuistiot/2014/t5lQ2u0cf/trm_2_2014_Saamelaisten_osallistumisoikeuksien_lisaaminen_valtion_maa-_ja_vesialueiden_kayttoa_koskevassa_paatoksentekomenettelyssa_saamelaisten_kotiseutualueella_.pdf> accessed 20 August 2014.

Art. 38.

[108] UN Human Rights Committee, Concluding observations on Belize in the absence of a report, adopted by the Committee at its 107th session (11–28 March 2013), 26 April 2013, CCPR/C/BLZ/CO/1. Para 25.

to cases involving relocation of indigenous peoples."[109] The jurisprudence of the Inter-American Human Rights Court and Commission will be dealt in the following chapter.

It should be emphasized that the concluding observations and general comments of the human rights monitoring bodies are not strictly legally binding.[110] They are interpretations of how a State Party should apply a treaty in order to fulfill its international obligations. According to the Committee of the International Law Association, however, treaty interpretation by monitoring bodies becomes authoritative if states do not oppose them.[111]

4.2 *FPIC in Jurisprudence of the Human Rights Monitoring Bodies*

The Inter-American human rights system has succeeded to create a rather expansive approach related to the rights of indigenous peoples. The recognition of FPIC has also found its way to both Inter-American Human Rights Commission and the Court.

In 2004, the Inter-American Commission on Human Rights expressed the requirement for FPIC in very clear terms in the case concerning the *Maya Indigenous Community v. Belize*, which concerned Maya land rights in their traditional territories in the south of Belize where the government had granted oil exploration and logging concessions.[112] The Commission reaffirmed the principle of consent by stating that granting such concessions "without effective consultations with and the informed consent of the Maya people" constituted a violation of their human rights.[113] The Commission importantly reaffirmed that international law upholds indigenous peoples' land and resource rights, independent of domestic law.[114]

The Inter-American Court of Human Rights has also developed an advanced protection of the collective rights of indigenous peoples, with the recognition of the

[109] Report on the Situation of Human Rights of a Segment of the Nicaraguan Population of Miskito Origin, OEA/Ser.L/V/II.62, doc.26 (1984), 120.

[110] See, generally, O'Flaherty, M. (2006). The Concluding Observations of United Nations Human Rights Treaty Bodies. Human Rights Law Review, 27, at 33.

[111] For an analysis on this issue, see the study of the International Human Rights Law and Practice – Committee of International Law Association (ILA), "Final Report on the Impacts of Findings of the United Nations Human Rights Treaty Bodies", www.ila-hq.org/html/layout_committee.htm (accessed 19 January 2014).

[112] *Maya indigenous community of the Toledo District v. Belize*, **Case 12.053, Report No. 40/04, Inter-Am. C.H.R., OEA/Ser.L/V/II.122 Doc. 5 rev. 1 at 727 (2004).** http://www1.umn.edu/humanrts/cases/40-04.html

[113] Para 194.

[114] Para 142. See also *Mary and Carrie Dann v. U.S.*, Case no 11.140, Report No. 75/02, Inter-Am. C.H.R., OEA/Ser.L/V/II.117, doc. 1 rev. 1 (2003), where the consent is recognized in Para 131.

FPIC. In its final ruling concerning *the case of Awas Tigni v. Nicaragua*,[115] the Court reaffirmed the principle that indigenous peoples have collective rights to their traditional lands and resources. The Court also held that these rights arise autonomously under international law.[116] The Court did not use the word "consent" in this case, but held, however, that the community's right to its own property prevent the Nicaraguan Government from unilaterally exploiting community natural resources.[117]

Thus, as pointed out by Anaya, the Court affirmed not only a right against state interference with indigenous peoples' rights to lands and resources without their consent, but also an affirmative right to state protection from such interference by third parties.[118]

Page argues that the Inter-American Court's emphasis of the full participation of the Awas Tingni community in the demarcation, and taking into account community's customary law, values and practices indicates the central role played in its decision by the principle of self-determination.[119] He continues his argument by stating that under the Court's interpretation, the American Convention on Human Rights protects indigenous communities' rights to property such that the right of each community to govern itself and to collectively organize its landholding is also protected. According to Page, the dual concepts of collective rights and self-determination for indigenous peoples are essential in understanding how FPIC may be properly implemented. Because the community as a whole must decide how it is governed, consent must also come from the community as a whole.[120]

Page continues by arguing that the proper implementation of FPIC will require a solid understanding of indigenous peoples' right to self-determination and cultural integrity, as well as to property and equality. There must be a firm basis in the customary laws and practices of the indigenous people concerned. Indigenous peoples must determine the standards by which to gauge whether consent is sought from a legitimate authority within their communities and whether conditions are such that their consent is in fact free and informed.[121]

The Inter-American Human Rights Commission and the Court are well known for their innovative approach that directly accounts for development in the United Nations human rights system. In the quite recent *Saramaka v. Suriname* case,[122] which surfaced after the adoption of the UN Declaration, the Inter-American Court utilized both, the Declaration, as well as common Article 1 of CCPR and CESCR

[115] The Mayagna (Sumo) Awas Tingni Community v. Nicaragua, Judgment of August 31, 2001, Inter-Am. Ct. H.R., (Ser. C) No. 79 (2001).

[116] Para 163.

[117] Para 164.

[118] Anaya, J. (2005), *supra* note 3, at 14.

[119] Page, A. (2004) Indigenous Peoples' Free Prior and Informed Consent in the Inter-American Human Rights System. Sustainable Development Law & Policy, 4(2), 16–20, at 16.

[120] *Ibid*, at 17.

[121] Page, A. (2004), supra note 118, at 19.

[122] *Saramaka People v. Suriname*, Inter-American Court of Human Rights, Judgment of November 28, 2007, Series C, No 172.

(the right to self-determination of peoples) as guidelines in adopting the concept of FPIC, as well as in interpreting the right to property in light of the right to self-determination of peoples.

In relation to logging and mining activities that had taken place in the territory of the Saramaka community, the Court made a special reference to Article 32 of the UN Declaration on the Rights of Indigenous Peoples, which requires states' consultation and cooperation with indigenous peoples in obtaining their FPIC prior to the approval of a project affecting their lands, territories, and other resources, as mentioned.[123]

In the *Saramaka* case, the Court importantly explained the exact content of the states' duty to consult indigenous peoples. According to the Court, consultations must be carried out via culturally appropriate procedures and with the objective of reaching an agreement. Furthermore, the Court continued by stating that the Saramakas must be consulted in accordance with their own traditions. This must not only take place when the need to obtain approval from the community arises, but during early stages of a development or investment plan. This is because early notice provides time for internal discussion within communities and for proper feedback for the state. The Court further noted that the state must ensure that members of the Saramaka people are aware of possible risks, including environmental and health risks, so that the proposed development or investment plan is knowingly or voluntarily accepted.[124]

Finally and significantly, the Court has noted that, in regard to large-scale development or investment projects that could have a major impact within the Saramaka territory, the state has a duty to not only to consult with the Saramakas, but to also obtain their FPIC based on their customs and traditions.[125] The Court ruled that the state shall adopt necessary legislative, administrative, and other measures in recognizing and ensuring the right of the Saramaka people to be effectively consulted.[126]

This case has a duple significance in relation to the concept of FPIC. Not only it creates a detailed requirement of FPIC in relation to large-scale interference in indigenous peoples' traditional lands and territories, but it also establishes a duty for States to obtain the consent of indigenous taking into account their customary laws and practices. The last remark is important while recognizing the attitude of true partnership between states and indigenous peoples in the sense that in the name of equality it must not be states but indigenous peoples setting rules for a dialogue on how FPIC be obtained in practice concerning each community in question. An additional importance of *Saramaka* case is that it endorses and directly uses the UN Declaration on the Rights of Indigenous Peoples as a source of legal interpretation. This case shows how so called "soft law" instruments can have a great legal weight, and it also demonstrates why some states have had difficulties to adopt the Declaration despite its non-binding nature.

[123] Paragraph 131 of the Decision.

[124] Para. 133.

[125] Para. 134.

[126] Paragraph 8 of the Operative Paragraphs.

In 2009, also the United Nations Human Rights Committee (HRC) made a historical shift by recognizing that the mere consultation of the indigenous community in question may not always satisfy the requirement of Article 27 of ICCPR when the culture and traditional way of life of indigenous peoples are seriously threatened. Until the case of *Poma Poma v. Peru,*[127] HRC had, in several cases, recognized that Article 27 requires effective participation and consultation of indigenous peoples in the matters that are significant to them.[128] In the *Poma Poma* case, for the first time, the HRC endorsed the right to FPIC. The case itself concerned a dispute over the exploitation of natural water resources, which caused a direct and negative impact on the indigenous Aymara peoples' traditional means of subsistence – the raising of llamas and alpacas on which the Ayamara community depended.[129]

The Committee reiterated its earlier view, according to which the admissibility of measures, which *substantially* compromise or interfere with culturally significant economic activities, depends on whether community's members have had the opportunity to participate in the decision-making process and whether they will continue to benefit from their traditional economy.[130] For the first time, in considering the meaning of the requirement of "effective" participation, the Committee stated that mere consultation is insufficient. Instead, FPIC was required. Additionally, according to the Committee, "the measures must respect the principle of proportionality so as not to endanger the very survival of the community and its members."[131] In the preset case, as pointed out by the Committee, there had been a complete lack of the consultation.[132]

The Committee concluded, based on the above mentioned facts and added with the fact that in this case the author, Ángela Poma, had been unable to continue benefiting from her traditional economic activity, that the activities carried out by the State party violated the right of the author to enjoy her own culture together with other members of her group, in accordance with Article 27 of ICCPR.[133]

---

[127] Human Rights Committee, Communication No. 1457/2006, Doc. CCPR/C/95/D/1457/2006 of 27 March 2009.

[128] See, for instance, I. Länsman et al v. Finland, Communication No. 511/1992; J. Länsman et al v. Finland, Communication no 671/1995, Apirana Mahuika et al v. New Zealand, Communication No. 547/1993, Lubicon Lake Band v. Canada, Communication No. 167/1984.

[129] Due to the building of wells, water had been diverted from the Peruvian highlands to a coastal city with the result that Aymara community living in the highlands had been deprived of their access to underground springs. The lack of water seriously affected the only means of subsistence of the community. For an analysis of the case, See K. Göcke (2010). The Case of Ángela Poma Poma v. Peru before the Human Rights Committee, The Concept of Free, Prior and Informed Consent and the Application of the International Covenant on Civil and Political Rights to the Protection and Promotion of Indigenous Peoples' Rights. Max Planck Yearbook of United Nations Law, 14, 337–370.

[130] *Ibid.*, at 7.6.

[131] *Ibid.*

[132] *Ibid.*, at 7.7.

[133] *Ibid.*

This case shows how the HRC is ready to expand the interpretation of Article 27 further than before, when the environmental interference on the lands of indigenous peoples is severe enough, and when the state has not committed to the profound consultation with the indigenous community. In viewing the Committee's comments, one may conclude that it is not so much the substantive protection that added anything new. The denial of the substantial harm had already been there. Indeed, only rarely has the Committee concluded that substantial harm has taken place. For this reason, alone, this case brings an additional weight to the protection of the cultural and natural resource rights of indigenous peoples under Article 27. Besides, and more importantly, the Committee expanded the scope of "effective" participation. It is clearly not a co-incidence that this decision was released shortly after the adoption of the UN Declaration on the Rights of Indigenous Peoples,[134] which endorses the concept of FPIC as stated before. Although the lack of a direct reference in the case to the UN Declaration has been criticized,[135] it is evident that the Declaration has played a role in this fundamental shift. This case is of particular significance since it shows that the FPIC has made its way not only to regional, but also in universally accepted human rights system. The future will show how consistent the case practice of the HRC will be in relation to the FPIC. As maintained, in the concluding observations, the Committee seems to shift between consultation and consent.

Although much still needs to be done in the implementation of FPIC, it should be noted that FPIC has found its way also in the human rights policies of industrial agencies. Although the World Bank still prefers the concept of free, prior and informed consultation, instead of consent,[136] many other development banks and agencies endorse this new principle and right, particularly after the adoption of the UN Declaration on the Rights of Indigenous Peoples.[137]

---

[134] After preparations lasting more than a decade, the UN Declaration on the Rights of Indigenous Peoples was adopted by the UN General Assembly on 7 September 2007. UN Declaration on the Rights of Indigenous Peoples, 7 September 2007, Sixty-first Session, A/61/L.67, http://www.iwgia.org/sw248.asp (accessed January 5, 2008).

[135] See Göcke, K. (2010), *supra* note 129, at 353–357.

[136] While the World Bank does not require FPIC, its new policy on indigenous peoples, OP 4.10 of 10 May 2005, requires obtaining indigenous peoples' broad community support through culturally appropriate and collective decision-making processes subsequent to meaningful and good faith consultation and "informed participation" at each stage and throughout the life of the project. Without such support the Bank will not proceed with project processing. IBRD/IDA, Operational Policy 4.10 on Indigenous Peoples, 10 May 2004, para. 1, 6 (c) and 11.

[137] E/cn.4/Sub.2/2003/38/Rev.2., para 10 (c). Similar statements on FPIC have been made by UN Special Rapporteurs on indigenous land rights (E/CN.4/Sub.2/2001/21); treaties concluded between states and indigenous peoples (E/CN.4/Sub.2/AC.4/1998/CRP.1), and indigenous peoples' intellectual and cultural heritage (E/CN4/Sub.2/1993/28), as well as by the Commission on Human Rights' Special Rapporteur on situation of the human rights and fundamental freedoms of indigenous people (E/CN.4/2002/97). In preparation for its 3rd session, the UN PFII distributed a questionnaire to all UN system "Indigenous Peoples Focal Points" in order to gather information about how the principle of FPIC is understood and applied by United Nations programmes, funds, agencies. (E/C.19/2004/11, para 3). The UNDP, UNFPA, FAO, ILO, UNITAR, IFAD, OHCHR,

The Expert Mechanism on the rights of indigenous peoples has made an Advice No. 4 (2012) in indigenous peoples' right to participate in decision-making, with a focus on extractive industries.[138] It reminds, referring to the Report of the Special Rapporteur on indigenous peoples,[139] that the right of indigenous peoples to participate in decision-making in relation to extractive activities is not confined to situations where indigenous peoples have a state-recognized title to the lands, territories and resources on or near which the extractive activity is to take place.[140] The Special Rapporteur has also reminded that states must take full responsibility in ensuring that adequate consultation is undertaken to obtain consent. A state cannot delegate this responsibility, even where it engages third parties to assist in consultation mechanisms.[141] The consultation is the starting point for seeking the free, prior and informed consent of indigenous peoples. If the potential impact or impacts are quite minor, the requirement to seek the free, prior and informed consent of indigenous peoples may not necessarily be required. Nonetheless, "the objective of consultations should be to achieve agreement or consensus."[142]

Guiding Principles on Business and Human Rights has reminded that while the state is the primary bearer of duties under international human rights law, business enterprises also have a responsibility to respect human rights. This means that they should avoid infringing on the human rights of others and should address adverse human rights impacts with which they are involved.[143]

Importantly, the Advice No. 4 states that extractive enterprises, as well as States, must bear in mind that indigenous peoples have the right to determine their own representatives themselves in accordance with their own procedures.[144] Indigenous

WHO responded that, while they do not have an official, working definition of FPIC, they recognize it as being embedded in the human rights framework and maintained, while not without challenges, that they "to large extent implement [FPIC] on an ad-hoc basis in line with the general guidelines, legal instruments and principles through which they work." (ibid., para 7). Also the International Finance Corporation's Micro-Finance Exclusion List states that IFC funds may not be used to finance "Production or activities that impinge on the lands owned, or claimed under adjudication, by indigenous peoples, without full documented consent of such peoples." Procedure for Environmental and Social Review of Projects. International Finance Corporation, December 1887, at 36; UNDP and Indigenous Peoples: A Policy of Engagement, paras. 26–30 (2001); European Bank for Reconstruction and Development (EBRD) Environmental and Social Policy, issued May 2008, at 50; The Asian Bank, Safeguard Policy Statement, Second Draft, October 2008, at 11–12; International Finance Corporation (IFC), Performance Standards on Environmental and Social Sustainability, January 2012, Para 11.

[138] A/HRC/EMRIP/2012/12.

[139] Report of the Special Rapporteur on the situation of human rights and fundamental freedoms of indigenous people (A/HRC/12/34).

[140] A/HRC/EMRIP/2012/12, at 16, referring to the Report A/HRC 12/34 (ibid), Para. 44.

[141] Report of the Special Rapporteur on the rights of indigenous peoples on extractive industries operating within or near indigenous territories (A/HRC/18/35), Para 63.

[142] A/HRC/18/42, annex, Para 9.

[143] Guiding Principles on Business and Human Rights, principle 11.http://www.ohchr.org/documents/issues/business/A.HRC.17.31.pdf (accessed 19 January 2014).

[144] A/HRC/EMRIP/2012/2, Para 15.

peoples should be able to make clear with whom governments and extractive enterprises should consult and from whom to seek the consent.[145] Where there are conflicting views on the legitimate representatives and/or representative structures of an indigenous people, the group should establish its own appropriate procedures to determine with whom governments and extractive enterprises should consult and/or seek consent. If necessary and desired, indigenous peoples can seek outside, independent assistance, including financial, to determine disputes.[146]

5 Concluding Remarks

During the last couple of decades, a paradigm shift has taken place in international law from exclusive to inclusive protection of the rights of indigenous peoples. Indigenous peoples have played themselves a crucial role in pushing their status from the cautious recognition of the distinctiveness of their culture towards recognition of their legal subjectivity in international law. Indigenous peoples' strengthened international status culminated in the adoption of the UN Declaration on the Rights of Indigenous Peoples with its rapid and wide effects, despite the lack of the strictly non-juridical nature of the instrument.

Free, prior and informed consent (FPIC) is a manifestation of a new paradigm where at least an internal self-determination and strong influential powers of indigenous peoples are now being recognized if not yet always implemented. FPIC dwells from the unique characteristic of indigenous peoples' rights, which recognizes the collective element in traditional, individually oriented human rights setting. It dwells from indigenous peoples' special relationship to their lands and resources as a fundamental basis of their identity and culture and thus their very lives.

The inherent link between indigenous peoples' traditional livelihoods and the use of natural resources has opened up human rights monitoring bodies to recognize a connection between the right to culture and a peoples' right to self-determination. Without this connection, it is difficult to imagine that indigenous peoples would ever have gained recognition as "peoples" in international law context. It is exactly their way of life based connection to the natural resources and its inherent link to the right to self-determination that has shaped indigenous peoples' international status from objects towards legal subjectivity. As it has been pointed out in this chapter, indigenous peoples were protected originally as a part of minority protection regime. Although the right of self-determination of indigenous peoples has been recognized in the UN-Declaration, we should not be misled by its nature: despite of many

[145] *Ibid.*, Para 16. The Special Rapporteur on the rights of indigenous peoples states: "Indigenous peoples may also need to develop or revise their own institutions, through their own decision-making procedures, in order to set up representative structures to facilitate the consultation processes." (A/HRC/18/35, Para 52).

[146] A/HRC/EMRIP/12/2, Para 16.

claims, in the state practice it still remains as a *sui generis* right, not guaranteeing an external self-determination in its fullest sense. Hence, it can be argued that indigenous peoples' international status is somewhere between that of minorities and peoples: indigenous peoples have become "semi-subjects" of international law.

FPIC has been seen as an important part of self-determination. It is indeed a right to have control over the issues that are fundamental to indigenous peoples as a collective. Yet, UN Human Rights Committee that accepts only individual claims, now endorses this concept as a part of the right to culture of individuals belonging to a minority group. Therefore, it is still somewhat unclear whether it can or should be argued that FPIC fundamentally rests mainly on the collective right of self-determination. Since the self-determination of indigenous peoples is still a very sensitive and disputed issue by many States, it might be a better idea to connect FPIC particularly to the right to cultural integrity or property, as the human rights monitoring bodies have done.

In fact, when we carefully study the birth of FPIC in the biodiversity protection instruments, we can see that its origin does not lay in the concept of self-determination. Instead, FPIC dwells from the idea that indigenous peoples, because of their traditional knowledge and practices, can make a valuable contribution to the maintenance of the biodiversity and, thus, should be taken into account in the decision-making as effectively as possible. Bio-cultural rights have been born out of the idea that biological and cultural diversity, in the case of indigenous peoples, go hand in hand.[147]

Anaya's argument, according to which it has become a generally accepted principle in international law that indigenous peoples should be consulted as to any decision affecting them,[148] seems quite right. He extends his argument by maintaining that the widespread acceptance of the norm of consultation demonstrates that it has become part of customary international law.[149]

Regarding the current legal status of FPIC, Ward rightfully argues that although a customary international legal principle that addresses indigenous peoples' full right to FPIC does not yet exist, there is a clear consensus within international human rights jurisprudence that, at a minimum, states must engage in good faith consultation with indigenous peoples prior to the exploration or exploitation of resources within their lands or actions that would impact their traditionally used resources.[150] Besides, as discussed in this chapter, two important human rights cases demonstrate that in cases where interference in indigenous peoples' traditional lands may cause significant, large-scale harm, mere consultation in good faith no longer is enough, but indeed FPIC is required. When studied the statements related to FPIC of different international actors, a conclusion can be made that at the time

[147] See, generally, Heinämäki and Herrmann (2013). The Recognition of Sacred Natural Sites of Arctic Indigenous Peoples as Part of their Cultural Integrity. Arctic Review on Law and Politics, 4(2), 206–231.

[148] Anaya, J. (2005), supra note 19, at 7.

[149] *Ibid.*

[150] Ward, T., *supra* note 50, at 54–55.

being, an unambiguous full embracing of the practical implementation of FPIC is still lacking: at the moment the most prevailing standpoint seems to support "consultation plus" model, which means that there has to be a real attempt to reach an agreement but, however, there is no real right to say no from the side of indigenous peoples. An indomitable attitude from the side of indigenous peoples is needed again to push for the final adoption of this new legal concept.

The UN Permanent Forum, together with other indigenous peoples' organizations and advocates keep on pushing an unequivocal interpretation and adoption of the right of FPIC in all decision-making concerning the rights of indigenous peoples. During the last few years, their voice has become louder and clearer. States, institutions and industrial agencies can no longer take their message lightly. It will only be a matter of time when FPIC finds its more wholesome way to national implementation. There, the Arctic States should take a leading role to implement this principle and right in national legislation and indigenous governance systems.

References

Anaya, J. (2004). *Indigenous peoples in international law* (2nd ed.). Oxford: Oxford University Press, at 63.

Anaya, J. (2005). Indigenous peoples' participatory rights in relation to decisions about natural resource extraction: The more fundamental issues of what rights indigenous peoples have in land and resources. *Arizona Journal of International and Comparative Law, 22*(1), at 9.

Anaya, J. (2009). The right of indigenous peoples to self-determination in the post-declaration era. In C. Chartres & T. Stavanhagen (Eds.), *Making the Declaration work: The United Nations Declaration on the Rights of Indigenous Peoples* (at 184). Copenhagen: IWGIA.

Baluarte, D. C. (2004). Balancing indigenous rights and a state's right to develop in Latin America: The Inter-American rights regime and ILO Convention 169. *Sustainable Development Law and Policy, 9*, at 10.

Barsh, R. L. (1994). Indigenous peoples in the 1990's: From object to subject of international law? *Harvard Human Rights Journal, 33*, reprinted In Watters, L. (2004) (Ed.). *Indigenous peoples, the environment and law* (pp. 15–42, at 23). North Carolina: Carolina Academic Press.

Barsh, R. L. (1996). Indigenous peoples and the UN Commission on Human Rights: A case of the immovable object and the irresistible force. *Human Rights Quarterly, 18*, 782–813.

Barsh, R. L. (2004). *supra* note 16, at 23.

Barsh, R. L. (2007). Indigenous peoples. In D. Bodansky, J. Brunnée, & E. Hey (Eds.), *The Oxford handbook of international environmental law* (pp. 830–831, at 839). Oxford: Oxford University Press.

Bartolome, C. (2005). The indigenous rights of participation and international development policies. *Arizona Journal of International and Comparative Law, 41*, at 41.

Colchester, M., & Mackay, F. (2004). In search of middle ground: Indigenous peoples, collective representation and the right to free, prior and informed consent. Forest Peoples Program. Retrieved January 19, 2014, from http://www.forestpeoples.org/sites/fpp/files/publication/2010/08/fpicipsaug04eng.pdf

Davis, M. (2008). Indigenous struggle in standard-setting: The United Nations Declaration on the Rights of Indigenous Peoples. *Melbourne Journal of International Law, 9*(2), 1–33, at 2.

Economic and Social Council Resolution. (1982)34. Retrieved January 19, 2014, from http://ap.ohchr.org/documents/E/ECOSOC/resolutions/E-RES-1982-34.doc

Fitzmaurice, M. (2009). The new developments regarding the Saami peoples of the North. *Journal on Minority and Group Rights, 16*, 67–156, at 151.
Göcke, K. (2010). The case of Ángela Poma Poma v. Peru before the Human Rights Committee, The concept of free, prior and informed consent and the application of the International covenant on civil and political rights to the protection and promotion of indigenous peoples' rights. *Max Planck Yearbook of United Nations Law, 14*, 337–370.
Heinämäki, L. (2011). Towards an equal partnership between indigenous peoples and states: Learning from Arctic experiences? *The Yearbook of Polar Law, 3*, 193–246.
Heinämäki, L., & Herrmann, T. M. (2013). The recognition of sacred natural sites of Arctic indigenous peoples as part of their cultural integrity. *Arctic Review on Law and Politics, 4*(2), 206–231.
Human Rights Commission Resolution (1982/19).
Koivurova, T. (2008). Alkuperäiskansojenitsemääräämisoikeuskansainvälisessäoikeudessa [The right of self-determination of indigenous peoples in international law]. In M. $. Aarto & M. Vartiainen (Eds.), *Oikeus kansainvälisessä maailmassa* [Law in a changing world] (pp. 249–269, at 268). Edita Publishing Oy, Lapin yliopiston oikeustieteiden tiedekunta [University of Lapland, Faculty of Law].
Koivurova, T., & Petrétei, A. (2014) Enacting a new mining act in Finland – How were Sami rights and interests taken into account? *Nordisk Miljörättslig Tidskrift, Nordic Environmental Law Journal*, 119–133. Retrieved September 1, 2014, from http://nordiskmiljoratt.se/onewebmedia/NMT%202014-1.pdf
Lubicon Lake Band v. Canada, communication No. 167/1984, CCPR/C/38/D/167/1884.
Magga, O. H. (2003, September 1). Presentation by the Chairperson of the Permanent Forum on Indigenous Issues, The UN Permanent Forum on Indigenous Issues – Ambitions and Limitations, Resource Centre for the Rights of Indigenous Peoples, Guovdageaidnu. Retrieved January 19, 2014, from http://www.galdu.org/web/index.php?artihkkal=39&giellal=eng
Page, A. (2004). Indigenous peoples' free prior and informed consent in the Inter-American Human Rights System. *Sustainable Development Law and Policy, 4*(2), 16–20, at 16.
Saramaka v. Suriname, Inter-American Court of Human Rights, Judgment of November 28, 2007, Series C, No 172, para 93.
Scheinin, M. (2000). The right to enjoy a distinct culture: Indigenous land and competing uses of land. In T. S. Orlin, A. Rosas, & M. Scheinin (Eds.), *The jurisprudence of human rights law: A comparative interpretive approach* (pp. 159–222, at 168). Turku/Åbo: Institute for Human Rights, ÅboAkademi University.
Seier, F. (2011). Free, prior and informed consent' under UNDRIP: What does it really mean? Right, Respect, Business and Human Rights Advisors. Retrieved January 19, 2014, from http://www. right2respect. com/2011/06/%E2%80%98free-prior-and-informed-consent%E2%80%99-under-the-un-declaration-on-the-rights-of-indigenous-peoples-what-does-it-really-mean
Stoll, P.-T., & von Hahn, A. (2005). Indigenous peoples, indigenous knowledge and indigenous resources in international law. In A. Xanthaki & N. Ghanea (Eds.), *Peoples and self-determination minorities* (pp. 3–14, at 33). Boston: Martinus Nijhoff Publishers.
Tauli-Corpuz, V. (2007). The Declaration on the Rights of Indigenous Peoples: A major victory and a challenge. Retrieved January 19, 2014, from http://www.twnside.org.sg/title2/resurgence/206/cover1.doc
The United Nations Declaration on the Rights of Indigenous Peoples, September 7, 2007, Sixty-first Session, A/61/L.67.
Ward, T. (2011). The right to free, prior and informed consent: Indigenous peoples' participation rights within international law. *Northwestern Journal of International Human Rights, 10*(2), 54.
Xanthaki, A. (2009). Indigenous rights in international law over the last 10 years and future developments. *Melbourne Journal of International Law, 10*, at 36.

Part V
Rhetoric and Protected Territories in the Arctic

Untouched and Uninhabited: Conflicting Canadian Rhetoric on the Protection of the Environment and Advancing Northern Economies

Katherine Sinclair

Abstract Positioning Arctic Canada as central to its political platform, the current federal government has set itself up to be one of Canada's most northern focused federal regimes in decades. This paper examines the ways that the Prime Minister's policy speeches portray and frame the Canadian Arctic's environment and land use, and the extent to which these statements incorporate broader ideas and premises about Arctic Canada. In particular, this involves an examination of the effects of policy portrayals in speeches of Arctic Canada as an untouched and uninhabited wilderness. This feeds into a tension between resource extraction in "isolated" areas as justifiable and the impulse to protect pristine places. This tension is significant, as catering to two very different expectations of land use places the federal government in a difficult position. Furthermore, emphasis on these land use plans may lead to a potential gap between federal government actions and the interests of northern residents, leading to a possible exclusion of northern, and in particular Inuit, priorities.

Keywords Canadian Arctic land use policy • Northern protected areas • Northern resource extraction • Policy communication • Inuit political exclusion

1 Introduction

This paper discusses Prime Minister Stephen Harper's political speeches about ArcticCanada as a way of communicating contemporary Government of Canada policies, strategies, and actions in the Canadian Arctic. It addresses the ways that these policy speeches portray and frame the Canadian Arctic environment and extractive industry activity, and the extent to which these statements incorporate broader ideas and premises about Arctic Canada. In particular, this involves an

K. Sinclair (✉)
Department of Anthropology, McGill University, Montréal, QC H3A 0G4, Canada
e-mail: katherine.sinclair@mail.mcgill.ca

T.M. Herrmann, T. Martin (eds.), *Indigenous Peoples' Governance of Land and Protected Territories in the Arctic*, DOI 10.1007/978-3-319-25035-9_12

examination of the effects of policy portrayals in speeches of Arctic Canada as an untouched and uninhabited wilderness. This feeds into a tension between resource extraction in "isolated" areas as justifiable and the impulse to protect pristine places.

By simultaneously advancing these differing land use strategies, the federal government is able to put forward two policy positions that may conflict with each other: the promotion of protected areas, giving the impression of a regime concerned about protecting the environment; and an emphasis on resource extraction, an economically driven action that seems to benefit northern residents by providing jobs, and southern Canadian residents by providing energy and resource revenues. Despite the tensions between these two approaches, advocating for both allows the Government of Canada to appear to be both a "hands off" government focused on the private industry stimulating the economy through resource extraction, as well as a "hands on" government interested in investing public funds in lands accessible to all Canadians.

2 The Government of Canada's Northern Resource Extraction Policies as Communicated through the Prime Minister's Speeches

This section draws on the actions of the federal government as communicated by Harper's speeches as well as the Canadian government's Northern Strategy, which is founded on four pillars: asserting Canada's sovereignty in the Arctic; boosting social and economic development throughout the Canadian north (especially through resourceextraction); protecting the northern environment; and, providing northern residents, including aboriginal people, with more control over their own affairs. The resource extraction-based governmental actions and policies focus on streamlining and advancing resource extraction in northern Canada, while the federal government's conservation initiatives increase the numbers of parks in Arctic Canada. Both are given weight in the Prime Minister's speeches, and both draw on the idea of the Arctic as a "frontier", "wilderness", and "natural". Here, discourse is used to further an agenda that tries to reconcile tensions between resource extraction and environmental stewardship, aiming to appear both aggressive in resource extraction as well as cognizant of environmental concerns. How is the tension between these differing policy approaches reconciled?

The current federal government has been characterized by some Canadian media as a government with a seeming lack of concern for the environment and a focus on encouraging subsurface resource extraction companies. This approach has been realized through a reduction of environmental oversight and consultations for extractive projects and cuts to previously established environmental projects. For example, in 2011 the Government of Canada withdrew from the Kyoto Accord, and in 2012 the federal government was criticized for, among other things, its cuts to environmental protection and an expansion of $6 billion in tax cuts to oil companies, banks, and corporations. The 2012 Bill C-38, the omnibus budget, included over $160 million in cuts to environmental spending, removed the requirements for environmental assessments of offshore drilling, and eliminated the Centre for Offshore Oil, Gas, and Energy Research. Bill C-38 also removed funding from the Experimental Lakes Area, a freshwater research center known as a global leader in research about household pollutants. Finally, the Bill also eliminated the National Round Table on the Environment and the Economy, an environmental watchdog agency that has been federally funded since its creation in 1988. Since many of these and other cuts were part of the omnibus budget, the relevant parliamentary committees did not evaluate them, and information about them was withheld from potential critics such as opposition Members of Parliament (MPs) (Wyld 2012).

The Canadian Northern Economic Development Agency (CanNor), and its Northern Projects Management Office (NPMO) in particular, aim in part to coordinate the work of the federal regulatory departments and agencies during the environmental review and permit phases of northern development projects. This is intended to streamline and speed up the approval process for resource extraction projects. While CanNor has many additional roles, such as research and advocacy, Prime Minister Harper focuses on the role of CanNor in resource extraction in his policyspeeches. Similarly, in 2008 Harper announced a $100 million investment over 5 years in the Geo-Mapping for Northern Energy and Minerals Program. This program aims to help the development of northern resources by charting the locations of mineral and energy resources. The federal government proclaimed an expected 5:1 return on this investment, projecting an estimated $500 million private sector investment as a result of the discovery of the location of these resources.

The announcements about these two specific projects, the creation and operations of CanNor and the Geo-Mapping Program, were paired with evocative descriptions of the Canadian North: The North was said to have "*potential*";

> Ladies and gentlemen, I'm very pleased to announce that our Government is launching the new geo-mapping for energy and mineral resources program. It will use state of the art geological science and technology to map the energy and mineral potential of the North. **Prime Minister Harper announces the Geo-mapping for Northern Energy and Minerals Program**, August 26, 2008, Ottawa, Ontario.
>
> On Tuesday I announced the creation of CanNor, the Canadian Northern Economic Development Agency, a new entity dedicated to helping the North realize its full potential. And an integral component of this agency will be the Northern Projects Management Office, a new initiative that will make the development review and approval process faster and more responsive to local interests and concerns. **PM announces labour market agreements and highway construction projects forCanada's northern Territories**, August 20, 2009, Yellowknife, Northwest Territories.

Subsurface natural resources were described as hidden "*treasure*";

> This information will be used to create geological models of the Arctic, subterranean maps that will help future resource producers find the treasures buried there. **Prime Minister Harper announces the Geo-mapping for Northern Energy and Minerals Program**, August 26, 2008, Ottawa, Ontario.

Northern natural resources were described as "*abundant*";

Few regions of the world are so richly endowed with natural resources as we are right here. **PM announces labour market agreements and highway construction projects forCanada's northern Territories**, August 20, 2009, Yellowknife, Northwest Territories.

And the northern landscape was said to be "*vast*", with connotations of vast here being large and unknown.

For instance, it's been said that finding minerals is like looking for a needle in a haystack, especially in a land so vast and challenging as Canada's North. Ultimately, industry has to find those needles, but the government can at least point it to the haystacks. That's why we pledge to produce new geological maps of the North, to help prospectors find energy and minerals. In fact, much of the fieldwork has now been done, and the results made available. Exploration companies are now using the data to decide where they will invest. For example, following public release of data from our energy and minerals geomapping program, large parts of Southwest Yukon were staked. **PM delivers remarks at Minto Mine, YK**, August 21, 2012, Minto, Yukon.

These announcements can be contextualized by the Prime Minister's other speeches about the North that mention resource extraction. The importance of resource extraction to the Government of Canada's Northern Strategy is indicated by the large number of times it is brought up in speeches on different issues. These speeches set the mental landscape in which specific program, policy, and strategy announcements are made. To the above characterizations we can then add the following descriptions: "*frontier*", connoting an area beyond the known borders that is large and unknown;

To that I say, government's first obligation is to defend the territorial integrity of its borders. And this will become more important in the decades to come – because northern oil and gas, minerals and other resources of the northern frontier will become ever more valuable. **Securing Canadian sovereignty in theArctic**, August 12, 2006, Iqaluit, Nunavut.

More and more, as global commerce routes chart a path to Canada's North – and as the oil, gas and minerals of this frontier become more valuable – northern resource development will grow ever more critical to our nation. **Prime Minister StephenHarperannounces new Arctic offshore patrol ships**, July 9, 2007, Esquimalt, British Columbia.

I have been to Whitehorse in Yukon, where settled life was established generations ago. To Iqaluit, Canada's newest capital city. To Jericho, the frontier diamond mine. And to Alert, on the very frontier of human existence. **The Call of the North**, *Address by the Prime Minister StephenHarper*, August 17, 2006 Yellowknife, Northwest Territories.

Just as the new Confederation looked to securing the Western shore, Canada must now look north to the next frontier – the vast expanse of the Arctic. **Prime Minister Stephen Harper announces newArcticoffshore patrol ships**, July 9, 2007, Esquimalt, British Columbia.

"*Progress*" (as a result of extraction);

So we ensure Northerners and all Canadians benefit from the jobs, prosperity and progress that flow from resource development. **The Call of the North**, *Address by the Prime Minister Stephen Harper*, August 17, 2006, Yellowknife, Northwest Territories.

And "*vast*" again, both in terms of land and, echoing that, the quantity of resources;

As Northerners, you know what I mean when I say that as you look out over the land, it seems endless, and so too are the North's possibilities. On Tuesday, I spoke of the vast

> mineral wealth upon with the economic future and orderly development of the North so vitally depends. **PM's announcement in Cambridge Bay**, August 23, 2012, Cambridge Bay, Nunavut.
>
> We are working with our territorial partners to develop the region's vast natural resources – to create jobs and prosperity for the benefit of Northerners and all Canadians. **The Call of the North**, *Address by the Prime Minister StephenHarper*, August 17, 2006, Yellowknife, Northwest Territories.
>
> We recognize the North is a vast storehouse of energy and mineral resources. We know that climate change is increasing accessibility to its treasures. **Prime Minister announces expansion of Canadian Forces facilities and operations in the Arctic**, August 10, 2007, Resolute Bay, Nunavut.
>
> This week I've made three announcements that underscore our commitment to economic development and environmental protection in the North. Just before leaving Ottawa on Tuesday, I announced the Geo-mapping for Energy and Mineral Resources Program. This program will map the geology of our northern territories to help prospectors and producers find the vast stores of gas, oil, gold, diamonds and other wealth buried beneath the tundra. **Prime Minister Harper announces the John G. Diefenbaker icebreaker project**, August 28, 2008 Inuvik, Northwest Territories.

These speeches and announcements about the federal government's action around resource extraction in the Canadian North therefore draw on the notions of the North as a resource frontier, as a vast area filled with a vast amount of resources or treasures, which can unlock the potential of the North that can result in Northern progress. In particular, this creates an image of the landscape as large, unknown, and untouched, in need of the federal government's intervention through policies, programs, and strategies to map, in a sense discover, and then develop the natural subsurface resources.

3 The Co-existence of Conservation Initiatives

This image of the North as vast, untouched, unknown, and uninhabited is also exemplified in Harper's speeches about conservation initiatives, as in line with one of the four arms of the federal government's Northern Strategy. Changes to the Arctic Waters Pollution Prevention Act are framed in both environmental terms and sovereignty concerns. The changes to the Arctic Water Pollution Prevention Act aim in part to protect marine and coastal environments by changing from regulating the shipping zones within 100 nautical miles to 200 nautical miles.

Like the Prime Minister's speeches about resourceextraction, the announcements of the creation of national parks similarly rely on the rhetoric of vast and untouched wilderness. For example, the land is referred to as '*untamed*';

> From Iqaluit to Whitehorse to Alert, I have met some of the diverse Canadians who are our true Northerners. They are the stewards of this largely untamed, sometimes harsh and always magnificent land, a vast storehouse of energy and mineral riches, a precious reservoir of ecological and cultural treasures. **Prime Minister announces the expansion of Nahanni NationalParkReserve**, August 8, 2007, Fort Simpson, Northwest Territories.

"Unspoiled";

Now, ladies and gentlemen, several years ago my wife Laureen was in Kuujjuaq, but this is actually my first trip to the Ungava Peninsula. Indeed, as I've seen with all my northern trips, this is a spectacular environment. The unspoiled beauty of this region immediately strikes anyone, and the unspoiled beauty is matched only by the extraordinary opportunities that lay before the people who live here. **Prime MinisterHarperhails landclaims agreement for Northern QuebecInuit**, March 28, 2008, Kuujjuaq, Nunavik.

These measures will send a clear message to the world: Canada takes responsibility for environmental protection and enforcement in our Arctic waters. This magnificent and unspoiled ecological region is one for which we will demonstrate stewardship on behalf of our country, and indeed, all of humanity. **Prime MinisterHarperannounces measures to strengthen Canada's Arctic sovereignty and protection of the northern environment**, August 27, 2008, Tuktoyaktuk, Northwest Territories.

"Rugged" and *"beautiful"*;

By increasing opportunities we are also strengthening our sovereignty and in doing so, we are living up to our commitment to preserve and protect our Arctic, to ensure that this rugged and beautiful place will be a strong and vibrant part of our country for generations to come. **Partnering to upgrade Yukon's Mayo B hydro facility**, August 21, 2009, Whitehorse, Yukon.

Containing *"natural wonders"* in need of *"preservation"*;

This is our government's Northern Strategy at its fullest: exercising our sovereignty by bringing people to the North, creating jobs for Northerners, while preserving and enjoying the wonders of the Northern environment. I make no apologies if all of this sounds a little bit like I'm presenting a tourism commercial, because over six summer tours, my passion for the North has continued to grow. **PM visits site of new Kluane NationalParkVisitor Centre**, August 26, 2011 Haines Junction, Yukon.

And as being a *"vastwilderness"*.

Our North, its landscape, its people, and their way of life are iconic in the minds of Canadians, central to our identity as the true North, strong and free. Even Canadians who have never seen this vast and beautiful wilderness know in their hearts that it defines us... As Prime Minister, I'm proud that through our Northern investments, through our investments particularly in Northern national parks, we are helping to preserve this priceless inheritance. **PM delivers remarks in Norman Wells, NWT**, August 22, 2012, Norman Wells, Northwest Territories.

Responsible development of resources, the preservation of wilderness, opportunities and improvements in the lives of people and their communities, the scientific quest for knowledge from the mountaintops to the sea beds... I, like many of you, am among those Canadians who have been fortunate enough to see so much of the wild and vast beauty of Canada's North. **PM delivers remarks in Churchill, Manitoba**, August 24, 2012, Churchill, Manitoba.

Like the speeches about resourceextraction, the statements on government initiatives to preserve northern landscapes draw on the notions of the North as an environmental frontier, a vast, untamed wilderness again filled with treasures (this time the natural environment, not the subsurface resources). This creates an image of the Northern land area as large, unknown, and untouched, which needs the preservationist intervention of federal government policies, programs, and strategies to be put to best use- an approach that contradicts, to some degree, the emphasis on potential destruction of land through resource extraction.

In addition to the contrast between extraction and protection, there is a further tension worth mentioning in the governmental policy approaches to northern land use: the role and position of the government. With respect to resource extraction, the federal government, via the Prime Minister's speeches, argues that an open, competitive market economy is needed:

> In a volatile, unpredictable world, investors are looking for stable, reliable producers of these commodities. Places where governments know what they want, where they understand wealth production is based on mutual benefit, and where government policies are founded on free-market economic principles, not self-serving political strategies.
>
> If Canada is to compete successfully for international investment capital [in resource development], these principles must be respected by all levels of government; national, provincial-territorial and local…We are committed to renewing and strengthening Territorial Formula Financing and Equalization. And a new deal on resource revenue-sharing is inseparable from those negotiations. But let me remind you again: It won't happen unless the North builds an open, competitive market economy where the resource industry can flourish. **The Call of the North**, *Address by the Prime Minister StephenHarper*, August 17, 2006, Yellowknife, Northwest Territories.

In comparison, when speaking of land preservation, the federal government advocates that governmental intervention in land use is a positive act that will benefit contemporary and future Canadians. Here, land use and the protection of land from development is the role of the government, not private industry:

> Ladies and gentlemen, preserving the region's ecological treasures is a key part of our plan for Canada's North. This is an important step in the process to ensure all the wonders of Nahanni will be protected so future generations can enjoy and appreciate them as we do today. Canada's New Government is committed to the preservation and protection of Canada's natural beauty. **Prime Minister announces the expansion of Nahanni NationalParkReserve**, August 8, 2007, Fort Simpson, Northwest Territories.
>
> These measures [as related to the Canada Shipping Act] will send a clear message to the world: Canada takes responsibility for environmental protection and enforcement in our Arctic waters. This magnificent and unspoiled ecological region is one for which we will demonstrate stewardship on behalf of our country, and indeed, all of humanity. **Prime Minister Harper announces measures to strengthen Canada's Arctic sovereignty and protection of the northern environment**, August 27, 2008, Tuktoyaktuk, Northwest Territories.
>
> We will protect the unique and fragile Arctic ecosystem for the generations yet to come. **Prime MinisterHarperannounces the John G. Diefenbaker icebreaker project**, August 28, 2008, Inuvik, Northwest Territories.
>
> So today I'm very pleased to announce that our government is establishing Canada's 44th national park, the Nááts'ihch'oh National Park Reserve… When combined with the Nahanni reserve, almost 35,000 square kilometers of land will be protected. With this announcement, we're continuing to move forward on our commitment to expand the national park system in the North. And we are protecting our environmental heritage for generations to come. As Prime Minister, I'm proud that through our Northern investments, through our investments particularly in Northern national parks, we are helping to preserve this priceless inheritance. **PM delivers remarks in Norman Wells, NWT**, August 22, 2012, Norman Wells, Northwest Territories.

These two different approaches to land use correspond to two very different styles of governance. The first represents a "hands off" approach to resourceextraction, encouraging a free-market economic plan. The second is a "hands on" approach to the creation of protected areas, giving the impression of a government invested in conserving the land from private exploitation for public use. Therefore, in addition to a tension between different types of land use, these two different styles of governance put forward by the Government of Canada in their Northern Strategy may also come into conflict with each other.

4 Tensions and Resolutions

There is a tension between these two important mandates of resource extraction and the preservation of northern landscapes that are the foci of government action in northern Canada. One involves arguably extensive destruction of land, while the other aims to preserve the landscape as it currently is. How can the tensions between these two different policy approaches be reconciled? How can the land be both exploited and used for resources, and untouched and preserved? These conflicting forms of land use illustrate a difficult position that the current Northern Strategy places the federal government in, needing to cater to two very different expectations of land use.

There is one further tension present in and beyond the speeches: the potential gap between federal government priorities and the interests of northern residents. What is lacking in the federal government's discourse and policyspeeches about resource-extraction and environmental stewardship are northern people.[1]Wilderness is emphasized at the expense of homeland, "untouched" at the expense of past and continued use, and southern-based policies at the expense of devolution and increased northern control over land and resources. In this way, reducing land use policy to either extraction or conservation precludes an emphasis on other forms of land use. Both types of land use advocated by the federal government run the risk of excluding northern residents, in particular Inuit, from land use for their own needs, including hunting and harvesting, via either sequestering land for extraction (and damage to wildlife and the environment that may result), or possible land use restrictions through the creation of protected areas. Proposals for and decisions about the management of different types of land use would most likely best address the interests of northern residents if they emerged from the North. The continued emphasis on policies emerging out of southern Canada for northern Canadian Territories may therefore exclude, rather than promote, the interests and priorities of the people who live there.

[1] It is beyond the scope of this paper to address the ways in which northern residents broadly, and Inuit specifically, are "talking back" to policies, in part via film, social media, photography, publications, prints, and more.

References

Harper, S. (2006–Present). *Selected public speeches*. Retrieved May 1, 2013, from http://www.pm.gc.ca/eng/media.asp?category=2

Published under the authority of the Minister of Indian Affairs and Northern Development and Federal Interlocutor for Métis and Non-Status Indians. (2009). *Canada's Northern Strategy*. Retrieved March 15, 2013, from http://www.northernstrategy.gc.ca/cns/cns-eng.asp

Wyld, A. (2012, December 31). A bleak year for environmental policy. *The Toronto Star*. Retrieved April 10, 2013, from http://www.thestar.com/opinion/editorials/2012/12/31/2012_a_bleak_year_for_environmental_policy.html

Index

T.M. Herrmann, T. Martin (eds.), *Indigenous Peoples' Governance of Land and Protected Territories in the Arctic*, DOI 10.1007/978-3-319-25035-9

The manufacturer's authorised representative in the EU is Springer Nature Customer Service Centre GmbH, Europaplatz 3, 69115 Heidelberg, Germany. If you have any concerns regarding our products, please contact ProductSafety@springernature.com

Printed and bound by CPI Group (UK) Ltd, Croydon, CR0 4YY
03/10/2026
02227506-0001